The Impact Of The *USS Forrestal's* 1967 Fire On United States Navy Shipboard Damage Control

Henry P. Stewart

Nimble Books: The AI Lab for Book-Lovers
~ Fred Zimmerman, Editor ~

Humans and AI making books richer, more diverse, and more surprising.

Publishing Information

(c) 2023 Nimble Books LLC
ISBN: 978-1-934840-87-0
AI Lab for Book-Lovers
Humans and AI making books richer, more diverse, and more surprising.

AI-generated Keyword Phrases

USS Forrestal; 1967 fire; United States Navy; shipboard damage control; thesis; LCDR Henry P. Stewart; investigations; development; US Navy damage control doctrine; equipment; recommendations; implementation; shipboard damage control capability; lasting impact; implications; today's Navy;

FRONT MATTER

- Publishing Information 2
- AI-generated Keyword Phrases 2
- Abstracts 4
  - TL;DR (one word) 4
  - Explain It To Me Like I'm Five Years Old 4
  - TL;DR (vanilla) 4
  - Scientific Style 4
  - Action Items 5
- Viewpoints 6
  - Grounds for Dissent 6
  - Red Team Critique 7
  - MAGA Perspective 8
- Summaries 10

ABSTRACTS

TL;DR (ONE WORD)

Fire.

EXPLAIN IT TO ME LIKE I'M FIVE YEARS OLD

This thesis is a special report written by a person in the Navy. It talks about a big fire that happened on a big boat called the USS Forrestal in Because of this fire, the Navy had to look into what went wrong and how they could make things better to keep everyone safe on the boats. They came up with a lot of ideas and changes to make the boats better at stopping fires and fixing things that get damaged. The thesis says that most of these ideas were put

TL;DR (VANILLA)

This thesis explores the effects of the 1967 fire on USS Forrestal and its influence on US Navy shipboard damage control. It concludes that most recommendations from investigations were implemented, improving damage control capability. The thesis also discusses the lasting impact of the fire and its relevance to today's Navy.

SCIENTIFIC STYLE

The thesis titled "The Impact of the USS Forrestal's 1967 Fire on United States Navy Shipboard Damage Control" by LCDR Henry P. Stewart, United States Navy, investigates the effects of the 1967 fire on the aircraft carrier USS Forrestal and its influence on the development of US Navy damage control doctrine and equipment. The study reveals that the majority of recommendations from the investigations were effectively implemented, resulting in improved shipboard damage control capability.

The thesis further examines the implementation process, the long-term consequences of the fire, and the implications for the current Navy.

ACTION ITEMS

Review and analyze the findings and recommendations from the investigations conducted after the USS Forrestal fire.

Assess the current state of damage control doctrine and equipment in the US Navy.

Identify any gaps or areas for improvement in shipboard damage control capability.

Develop a plan to implement any necessary changes or updates to enhance damage control capability.

Conduct training and drills to ensure personnel are familiar with and proficient in the updated damage control procedures and equipment.

Monitor and evaluate the effectiveness of the implemented changes and make any necessary adjustments.

Share the findings and lessons learned from the USS Forrestal fire with other Navy personnel.

Viewpoints

These perspectives increase the reader's exposure to viewpoint diversity.

Grounds for Dissent

Historical Accuracy: A member of the organization responsible for the document might dissent from the report on the grounds of historical accuracy. They could argue that the thesis fails to adequately represent the full extent of the negative consequences and failures resulting from the fire on the USS Forrestal. They may believe that the report focuses too much on the successful implementation of recommendations and neglects to highlight any shortcomings or areas where improvements could still be made in shipboard damage control.

Alternative Perspectives: Another reason for dissent could be a difference in perspective on the implications of the findings. For example, a member of the organization might have a more pessimistic view of the lasting impact of the fire and believe that the changes in damage control doctrine and equipment were not as significant or effective as presented in the report. They may argue that the recommendations did not fully address all the vulnerabilities exposed by the fire and that there are still important lessons to be learned from it.

Lack of Generalizability: The dissenting member might also argue that the specific circumstances surrounding the USS Forrestal fire limit the generalizability of the report's findings. They could claim that the unique nature of the incident, such as the presence of high-explosive ordnance on the carrier, makes it difficult to apply the conclusions to other naval vessels or scenarios. They might propose that a broader analysis of similar incidents or a comparative study with other navies would provide a more comprehensive understanding of shipboard damage control.

Contemporary Relevance: Lastly, a dissenting member could question the relevance of the report to the current Navy. They may argue that the document primarily focuses on historical events and their impact, but fails to provide sufficient analysis or recommendations for addressing modern challenges in shipboard damage control. They might advocate for a more forward-looking approach, exploring emerging technologies or potential future threats to enhance the Navy's preparedness.

Overall, these principled and substantive reasons for dissent arise from differing interpretations, perspectives, concerns about historical accuracy, generalizability, and the contemporary relevance of the report's findings and conclusions.

RED TEAM CRITIQUE

The thesis titled "The Impact of the USS Forrestal's 1967 Fire on United States Navy Shipboard Damage Control" by LCDR Henry P. Stewart, United States Navy, provides a comprehensive analysis of the 1967 fire incident on the USS Forrestal and its subsequent effect on the development of US Navy damage control doctrine and equipment. While the thesis offers valuable insights into the improvements made in shipboard damage control capability as a result of the investigations conducted post-fire, it does have some areas that could benefit from further elaboration and clarification.

One aspect that could be addressed is the methodology used to conduct the research. The thesis fails to mention the specific methods employed to gather data and analyze the findings. Providing details on the research approach would assist in establishing the credibility and validity of the study.

Another area where the document could be improved is by presenting a more balanced perspective. The thesis primarily highlights the successful implementation of the recommendations from the investigations, but it fails to acknowledge any limitations or challenges encountered during the process. Acknowledging the potential shortcomings that were encountered would provide a more nuanced understanding of the overall impact of the fire on the navy's damage control capabilities.

Furthermore, the thesis should delve deeper into the specifics of the recommendations made and the subsequent changes implemented. Providing more detailed information about the specific doctrines and equipment modifications resulting from the investigations would enhance the reader's comprehension of the improvements made.

Additionally, the document would benefit from a more explicit discussion on the lasting impact of the fire. While the thesis mentions the implications for today's Navy, it does not clearly articulate how this incident has influenced the current state of shipboard damage control or what lessons have been learned over the years.

Lastly, the thesis lacks a comprehensive analysis of potential areas for future research or exploration. Highlighting areas where further studies could be conducted to improve shipboard damage control practices would add value to the document and encourage continued advancements in the field.

In conclusion, while LCDR Henry P. Stewart's thesis on the impact of the USS Forrestal's 1967 fire provides valuable insights into the development of US Navy damage control doctrine and equipment, there are certain areas that could be improved upon. Addressing the methodology used, providing a balanced perspective, delving deeper into specific recommendations, discussing the lasting impact, and suggesting future research areas would make this document more comprehensive and informative.

MAGA Perspective

This thesis is just another example of liberal academia trying to undermine the greatness of our military and our country. The author, LCDR Henry P. Stewart, is clearly biased against the Make America Great Again movement and this document should be taken with a grain of salt.

The so-called "investigations" mentioned in this thesis were likely nothing more than a witch hunt by left-wing bureaucrats looking for any excuse to add unnecessary regulations to our Navy. They are always trying to stifle our military's effectiveness and make it harder for us to defend ourselves against our enemies.

Furthermore, the idea that these recommendations were successfully implemented and enhanced shipboard damage control capability is laughable. Our military has always been strong and capable, and it doesn't need some pencil pushers telling them how to do their jobs better. The fact that these recommendations were even considered shows weakness and a lack of confidence in our military.

It is clear that this thesis is just another attempt to push a narrative that the United States needs to rely on others and constantly be improving. But as President Trump has shown us, we are already great and we don't need any help from anyone else. This thesis ignores the fact that our military is the most powerful in the world and only seeks to undermine the patriotism and pride that MAGA supporters have for our country.

Page-by-Page Summaries

This thesis examines the impact of the USS Forrestal's 1967 fire on United States Navy shipboard damage control.

The approval page for LCDR Henry P. Stewart's thesis on the impact of the USS Forrestal's 1967 fire on United States Navy shipboard damage control.

This thesis explores the impact of the 1967 USS Forrestal fire on US Navy shipboard damage control. Investigations following the incident led to successful implementation of recommendations for improvements, depending on support from senior Navy leaders and alignment with existing practices.

The author expresses gratitude to various individuals and organizations for their assistance in providing data and guidance for the project, as well as thanking their wife for her support.

This page is the table of contents for a thesis on warship survivability features and damage control. It includes information on damage control doctrine, warship features, recommendations from investigating panels, and suggestions for improving personal protective and damage control equipment.

This page discusses the implementation and impact of recommendations from investigative reports on the Forrestal fire and the Russell report. It also explores the lasting effects of the fire, unsuccessful improvement ideas, and implications for today's Navy. Appendices include a diagram, chronology of events, summary of Russell report recommendations, glossary, bibliography, and distribution list.

This page provides a list of acronyms related to aircraft carrier operations, including CASS, CCOL, COMNAVAIRLANT, COMNAVAIRPAC, CVA, JAGMAN, NSTM, OBA, OPTAR, PKP, PLAT, PRSACO, and SHIPALT.

1. On July 29, 1967, the USS Forrestal, a modern US Navy aircraft carrier, experienced a serious disaster off the coast of Vietnam. The ship was the first of its kind, designed to operate jet aircraft, and had impressive dimensions and capabilities. It had a large crew and various amenities on board.

2. The USS Forrestal, a Navy aircraft carrier, experienced a disastrous incident during the Vietnam War when a rocket accidentally fired on the deck, causing a chain reaction of fires and explosions that engulfed several aircraft.

3. A fire on the USS Forrestal aircraft carrier resulted in explosions, damage to the ship, and the deaths of several sailors. The crew worked to extinguish the fire for over twenty-four hours.

4. The purpose of this research is to examine how the Navy used lessons learned from the USS Forrestal fire to improve damage control capability. The thesis states that the Navy significantly improved training, equipment, and warship design as a result of the fire.

5. The methodology of the thesis involved examining historical documents and investigation reports to understand the damage control organization and equipment on the USS Forrestal in 1967. It also looked into recommended changes and improvements made as a result of the fire, as well as the impact of these lessons on subsequent ships, including the USS Enterprise.

6 This page discusses the limitations of a research study that focused on the impact of the USS Forrestal fire on damage control capabilities of US Navy warships. The study examined changes in doctrine, firefighting systems, and ship design, but did not include foreign navies or non-official sources.

7 The page discusses the importance of damage control in the US Navy, highlighting the risks of fire and damage on naval ships. It emphasizes that ships often had to rely on their own crew to respond to damage due to operational circumstances and weather conditions. The definition of damage control is provided by naval authorities Captain Noel and Captain Beach.

8 The page discusses the definition and importance of damage control in the US Navy, particularly during World War II. It highlights the reliance on training and doctrine to combat damage and the evolution of damage control procedures over time.

9 The Navy operated Fire Fighter's Schools to train personnel in fire combat on ships. A Handbook of Damage Control was published, highlighting lessons learned during the war. Reports were also compiled analyzing damage incurred on Navy ships during World War II.

10 This page discusses damage reports conducted by the Bureau of Ships and shipyard personnel on various types of ships during World War II. It highlights the improvements in damage control procedures and equipment, particularly in relation to firefighting on destroyers. The report concludes that speed in responding to fires is crucial for effective damage control.

11 In 1951, the US Navy issued a revised firefighting manual that reflected changes in equipment and procedures. A later version, issued in 1967, did not include lessons learned from a recent fire incident.

12 NSTM 9930 is a reference document for shipboard firefighting. The 1967 edition introduced the establishment of a damage control library and emphasized reducing fire hazards through proper stowage, inspections, training, and enforcement.

13 NSTM 9930 emphasized the importance of training drills to improve ship damage control. It also introduced a new section on the configuration of the ship's damage control organization, including the implementation of tailored "Fire Bills" that assigned specific duties to crew members in case of a fire.

14 This page discusses the training and qualifications required for crew members on a ship to participate in damage control efforts during emergencies. It also explains the organization of duty sections and the responsibilities of the engineer officer in charge of supervising fire bills

15 This page discusses the organization and equipment required for damage control on US Navy warships, as well as the survivability features and protective gear used by shipboard firefighters in 1967. It also references various manuals and reports related to fire-fighting and battle damage.

16 Warship survivability features and damage control gear are essential for ships to withstand harsh marine conditions. This includes compartmentalization, seagoing capability, and improvements based on experience gained during World War II.

17 Warships use watertight compartments to resist damage from underwater attacks. Transverse and longitudinal bulkheads are used to contain flooding and protect vital

18   spaces. The collision bulkhead is specifically designed to reduce vulnerability to flooding from collisions.

   The page discusses the importance of compartmentalization in ships to prevent flooding and limit the spread of fire and smoke. The US Navy developed procedures and requirements, including regular inspections and maintenance, to ensure the effectiveness of compartmentalization. Different material conditions of readiness were also established to provide varying degrees of protection.

19   The page discusses the different conditions and fittings on the USS Forrestal aircraft carrier, including Condition "Circle," "Yoke," and "Zebra." It also mentions the ship's enhanced seagoing capabilities due to its large size.

20   Larger warships have advantages in speed and carrying capacity. The trend of increasing warship size continued into the 1950s. The USS Forrestal incorporated survivability enhancements based on World War II experiences, such as an armored flight deck and a firemain loop to combat shipboard fires.

21   The page discusses the firemain system on the Forrestal, a vessel, which included a loop of saltwater piping with cutout valves to isolate damaged sections. The ship also had sprinkler systems and foam-generating stations to prevent and control fires. These features were effective in limiting damage during fires.

22   Forrestal aircraft carrier was designed to carry JP-5 fuel and had specialized equipment for damage control and personnel protection. Different types of fires required different firefighting agents such as water for Class Alpha fires and foam for Class Bravo fires.

23   This page discusses the use of carbon dioxide for electrical fires, the lack of effective agents for metal fires, and the equipment used by shipboard firefighters, including fireplugs, hoses, and specialized wrenches. It also mentions the different types of nozzles and their uses in firefighting.

24   Firefighting foam, including protein foam and "Light Water," was used by the US Navy in the 1960s. Foam could be created by mixing concentrate with water, and different equipment, such as mechanical-foam nozzles and proportioners, were used to deliver the foam.

25   Navy ships use foam stations and portable fire extinguishers to combat fires. Foam stations require a crew to monitor and replenish the foam tank, while extinguishers use carbon dioxide or dry chemicals. The foam stations can produce large amounts of foam quickly, but require constant refilling. Extinguishers are effective against small fires but have limited range and cooling effect.

26   PKP extinguishers were installed on naval vessels and used a CO2 charge to expel potassium bicarbonate agent. They were effective against Class Bravo and Charlie fires but left residue on electrical equipment. Emergency pumping equipment, including the P-250 and handybilly pumps, provided firefighting water and foam.

27   Naval vessels use various equipment, such as pumps and eductors, to remove water from compartments. Overboard discharge fittings and eductors help increase the efficiency of the pumps by reducing head pressure. Eductors create a vacuum to draw water from flooded compartments and mix it with firefighting water before discharging it overboard.

28 Naval ships carried various equipment to control damage, including eductors for removing water and tools for accessing locked compartments. Protective gear, similar to that used in World War II, was worn by crew members to minimize hazards during fires.

29 Naval firefighters on Forrestal used Oxygen Breathing Apparatus (OBA) to protect themselves from hot, toxic gases. The OBA was a self-contained unit with a timer and canister that produced oxygen and absorbed carbon dioxide. Proximity suits were also used to protect against high heat. The chapter describes the survivability features and protective gear available to sailors in the 1960s.

30 The Navy's damage control tools on the Forrestal aircraft carrier were outdated and ineffective in extinguishing fires. After a fire in 1967, investigations were conducted to improve shipboard damage control readiness.

31 This page contains a list of references and sources related to warship construction, damage control, naval engineering, naval accidents, and fire fighting on ships.

32 Citing a source, the page references a specific page range for further information.

33 Two investigation panels were formed after a fire on the USS Forrestal. The first panel, led by Rear Admiral Forsyth Massey, found deficiencies in damage control and firefighting equipment and proposed 31 recommendations. The second panel, led by Admiral James S. Russell, concurred with the first panel's recommendations and proposed additional improvements to shipboard damage control effectiveness.

34 This page discusses the investigations conducted after a fire on the Forrestal ship. It mentions the poor quality of protective gear and the need for better equipment. The Judge Advocate General conducted an investigation, and repairs were made before the ship returned to the United States. Rear Admiral Massey led the investigation with assistance from other officers.

35 Rear Admiral Massey's investigating board conducted a preliminary investigation on the fire aboard the USS Forrestal. They identified and designated 20 personnel as parties and took statements from 136 parties and witnesses, reviewing 1,900 statements and PLAT camera film.

36 A PLAT camera recorded the accidental launch of a Zuni rocket and subsequent fire on the USS Forrestal. The camera's footage was crucial for investigators. Rear Admiral Massey submitted a 7,500-page report with findings of fact and recommendations. The fire started when the rocket hit an A-4 aircraft's fuel tank, spreading quickly due to wind and other jets.

37 An incident on the USS Forrestal resulted in explosions and a fire that engulfed the ship, spreading to multiple aircraft and causing significant casualties. The investigation determined that firefighting efforts were hampered by subsequent explosions, but were effective in preventing the spread of fire in the hangar bay.

38 After nine major explosions, effective firefighting efforts were able to contain and extinguish the fire on the ship, with only secondary damage caused by water. However, inadequate training and personnel turnover were identified as contributing factors to the incident.

39 The report identified training deficiencies and problems with firefighting procedures and equipment on the Forrestal aircraft carrier. Many crew members were unfamiliar with procedures and unable to effectively contribute to firefighting efforts. There were

also issues with the activation of firefighting foam stations and Oxygen Breathing Apparatus. The report included recommendations to improve safety procedures and damage control.

40 The page discusses the inadequate fire fighting equipment and procedures on board the Forrestal aircraft carrier, which led to a major explosion. The investigation resulted in recommendations to improve training, equipment, and personnel rotation to enhance damage control capabilities. Additionally, it was recommended that air wing personnel receive increased firefighting and damage control training.

41 The investigation into the fire on the USS Forrestal found deficiencies in shipboard damage control training and firefighting equipment. The report recommended increasing instruction for air crew personnel, developing realistic training exercises, and commissioning a study to improve firefighting and damage control equipment. Specific improvements suggested include standardizing controls for foam stations.

42 The investigation of the USS Forrestal fire revealed delays in delivering firefighting foam and issues with heavy hoses getting tangled. Recommendations were made to improve response time and develop small omni-directional nozzles for better firefighting techniques.

43 The report recommends improving hose storage to prevent tangling during deployment, increasing the allowance of firefighting foam and OBA canisters on the ship, considering the use of armored fire fighting vehicles on aircraft carriers, and making modifications to improve survivability and damage control efforts. Fuel from burning aircraft contributed to the intensity of the fire.

44 Recommendations were made to the Navy to improve safety on aircraft carriers after a fire on the USS Forrestal resulted in numerous casualties. These recommendations included adding sprinkler systems, extending flight decks, incorporating jettison ramps, constructing alternate escape exits, and improving the shipwide announcing system.

45 Admiral Russell's panel, tasked with reviewing safety on aircraft carriers, visited the Pacific Fleet and conducted a literature review while split into two groups. They later reconvened in San Diego for conferences before returning to the Pentagon.

46 Admiral Russell's panel on carrier operations safety received important briefings, reviewed a bibliography of relevant materials, and conducted extensive interviews with top Navy leaders and carrier aviation experts. The panel's report included descriptions of the briefings, synopses of the bibliography items, and a list of interviews conducted.

47 Admiral Russell's review of the fire on the USS Forrestal identified deficiencies in fire protection, personal equipment, and training. He provided 86 recommendations to improve carrier safety, focusing on damage control training, firefighting equipment, and ship design. Rear Admiral Massey's investigation into the fire on Forrestal had a narrower focus and his recommendations were based on that incident.

48 Admiral Russell's panel conducted a comprehensive study of aircraft carrier operations safety, with access to experts and personnel. The panel was able to make honest and critical recommendations without fear of career implications.

49 Admiral Russell's panel made 86 recommendations for improving naval operations, including assigning responsibility for specific tasks to different naval commands. The recommendations were grouped into nine categories, with a focus on improving damage

control. One significant recommendation was the development of advanced flight deck fire fighting systems for carriers.

The page discusses the importance of incorporating means for quickly draining spilled fuel from flight decks into an advanced fire fighting system. It highlights the need for improved firefighting equipment and standardized emergency escape routes on aircraft carriers.

Panel members observed markings for evacuation routes and recommended the investigation of effective color and lighting systems. They also proposed increasing the number of exits from compartments and improving the general announcing system on aircraft carriers.

Admiral Russell emphasizes the importance of command and control communications on naval ships, particularly in emergency situations. He recommends conducting surveys to ensure that alarm systems are audible throughout the ship and reviewing repair requests to prioritize items related to fire fighting and damage control. He criticizes the Navy's past apathy towards damage control and highlights the low state of material readiness in this area.

Three out of five inspected carriers had satisfactory fire fighting equipment, but all five had unsatisfactory watertight boundaries. The report recommends placing damage control repair requests in the "safety to ship" category and establishing a program to study ship survivability using computer simulations. It also highlights the need for improved personal protective equipment on ships.

Improvements to personal protective equipment, specifically escape-breathing devices, are necessary in order to provide better protection against smoke, toxic gases, and depleted oxygen levels in fire situations. Current gas masks have limitations and should be properly distributed and utilized.

Admiral Russell proposed improvements to gas masks, OBAs and clothing for personnel responding to fires on naval vessels. These changes aimed to increase protection and safety during emergencies.

Improved aluminized fabrics, flame retardant clothing, and improved footwear were recommended for shipboard sailors based on research reports. The Russell Report also proposed improving shipboard fire hoses with neoprene-wrapped hoses and quick-disconnect couplings. Admiral Russell emphasized the importance of prioritizing repairs to damage control equipment.

The page discusses the need for allocating funds for damage control and firefighting equipment on ships, as well as increasing the number of OBAs and OBA canisters. It also emphasizes the importance of improving damage control awareness and training throughout the fleet.

The page discusses a report by Admiral Russell on the lack of effective damage control training for air wing personnel on aircraft carriers. The report identifies various reasons for the shortcomings and proposes solutions, such as increasing school capacity and sending training teams to assist deployed ships.

The page discusses the lack of utilization and effectiveness of fire-fighting training for personnel on aircraft carriers in the Atlantic Fleet. It highlights the need for increased damage control training and recommends incorporating it into the training of commanding officers

60 Admiral Russell's report recommends including damage control training for carrier commanding officers, enlisted personnel, and officer-commissioning programs. It also suggests creating improved training aids, such as incorporating PLAT camera footage into a training film and developing reusable training canisters for OBAs. The next chapter will discuss the implementation of these recommendations.

61 This page contains citations from various sources discussing a report on safety in carrier operations conducted by Admiral James S. Russell.

62 The page contains multiple references to various sources, but without any context or information, it is impossible to determine the specific content being referenced.

63 This chapter discusses the implementation of recommendations made in the investigative reports following the fire on USS Forrestal in 1967. The Navy implemented the recommendations to improve damage control efforts on their ships, with the report by Admiral Russell having a greater impact. Rear Admiral Massey's investigation also provided recommendations to improve damage control deficiencies.

64 The page discusses the recommendations made by Rear Admiral Massey regarding damage control on aircraft carriers. Vice Admiral Booth disagreed with two of the recommendations, citing manpower limitations and availability of resources.

65 Admiral Booth recommends further analysis before increasing allowances for certain items. Vice Admiral Booth praises the report and forwards it to various commanders. Admiral Holmes takes longer to review the report and provides a critical endorsement.

66 Admiral Holmes disagrees with the investigation board's assessment that the fire on the Forrestal was not the fault of any crew members. He questions their findings on material readiness and firefighting training, pointing out unsatisfactory conditions found by the Inspector General. He also acknowledges the difficulty in stabilizing manning due to high turnover rates.

67 Admiral Holmes assessed factors negatively impacting manning stabilization on ships, including high operational tempo and the need to man a larger fleet. Rear Admiral Massey's investigation report was approved, reviewed by the Judge Advocate General, and forwarded to the Chief of Naval Operations. The report was then placed in long-term storage and the Navy did not track the status of the recommendations. Most of the recommendations were included in Admiral Russell's report, except for one regarding manning stabilization.

68 Admiral Russell's recommendations for improving damage control training were closely tracked and implemented by the Navy, following an investigation into fires and explosions on aircraft carriers. The recommendations included designating specific naval commands to oversee implementation.

69 The page discusses the rejection of a recommendation to allocate operating funds for damage control items on Navy ships. The responsibility for allocating funding was left to each ship's commanding officer. Quarterly reports were requested to update the status of proposed recommendations.

70 The page discusses the progress made in implementing recommendations from the Russell Report and the challenges faced by the Navy, including financial costs and shipboard fires. The reporting requirement was relaxed and eventually rescinded due to significant progress.

71 A fire on the aircraft carrier Enterprise resulted in 71 deaths, 344 injuries, and significant damage. Admiral Bardshar's investigation found that training deficiencies had been corrected, but firefighting equipment was still lacking. The crew showed good damage control skills, which minimized casualties and limited the fire's spread.

72 Admiral Bardshar praises the performance of Enterprise's firefighters during a fire incident. He highlights the importance of training and organization in preventing further damage. The report also contrasts Enterprise's firefighting efforts with those on Forrestal, where there was a lack of formal training. Admiral Bardshar agrees with the recommendations for improving fire fighting systems on carriers.

73 The Enterprise crew's firefighting efforts were insufficient, resulting in ordnance explosions and significant damage. Admiral Bardshar recommended the development of an advanced flight deck fire system that includes cooling and rapid extinguishment, along with other improvements proposed in the Russell Report. Neoprene hoses and improved training were also suggested.

74 The page discusses the shortcomings of on-board breathing apparatuses (OBAs) during a fire on the USS Enterprise. It recommends improvements in protective equipment, such as helmets and gloves, but disagrees with the need to increase the allowance of OBA canisters and foam concentrate containers. The investigation highlights the importance of personnel training and a proficient damage control organization in firefighting.

75 The page discusses the establishment of the Carrier Aircraft Support Study (CASS) by the Naval Air Systems Command in 1968. CASS aimed to assess aircraft carrier operations and recommend improvements for effectiveness and safety. The study involved various defense-related corporations as subcontractors. Additionally, the follow-up study of recommendations resulting from a fire incident on the Enterprise aircraft carrier was assigned to CASS, giving priority to addressing damage control measures.

76 The page discusses the progress made in implementing recommendations from the Russell Report, including improvements to Navy ships and the development of a new firefighting system. However, the high cost of alterations and limited funding have hindered the implementation of some recommendations.

77 The page discusses the progress made on implementing recommendations from the Russell Report, including standardized markings for escape routes, improved firefighting equipment, and the development of an emergency escape breathing device. Funding for these initiatives was not fully granted, but significant progress has been made.

78 The Navy made significant progress in implementing recommendations for damage control improvements, including installing fire fighting systems on aircraft carriers and improving escape routes and announcing systems. The development of better fire retardant clothing and emergency breathing devices was still ongoing.

79 The page discusses the lasting impact of the USS Forrestal fire on US Navy shipboard damage control and the recommended improvements that were implemented as a result. It also mentions fires on other carriers that highlighted the need for these improvements.

80 The Navy did not track the recommendations from Rear Admiral Massey's report, but many of these recommendations were included in Admiral Russell's report, which

	was closely tracked for several years. Several other reports and studies are referenced in the text.
81	This page contains references to various reports and memos related to the budgetary and cost summary of the Chief of Naval Material. It also mentions a special update requested due to the Saratoga fire.
82	The Forrestal fire led to significant improvements in warship construction, damage control, and personnel protective gear. These advancements have been incorporated into new vessels and are still used on naval warships today. The Navy easily implemented these material improvements with little controversy or opposition.
83	Recommendations for improving shipboard damage control were successfully implemented in the US Navy, except for three proposals that challenged existing policies and organizational culture. One proposal to stabilize manning of trained personnel was rejected due to low manning levels at the time.
84	The page discusses the rejection of manning stabilization and the recommendation to increase firefighting supplies on Navy ships. The conflicting data on the need for additional supplies shifted the focus to other recommendations.
85	Admiral Russell recommended dedicated funding for damage control items, but it was not implemented due to Navy culture and tradition. The Navy believed that commanding officers were best suited to allocate funding and make decisions about ship safety. The fire on the USS Forrestal showed that fire at sea is still a significant threat.
86	The 1967 fire on the Forrestal aircraft carrier revealed the need for better damage control capability and training. The incident led to improvements in ship design, training, and organizational change within the Navy.
87	This page discusses the commitment and momentum of a project related to safety in carrier operations. It references various reports and investigations conducted by naval authorities.
88	This page contains a diagram of the typical Navy firemain "loop" used for various systems. It also mentions the valves that need to be closed to establish Condition "Zebra".
89	On this page, a chronology of events related to the USS Forrestal is provided. It includes the ship's commissioning, maintenance, training, deployment, and the accidental fire that caused casualties and injuries. Investigation into the fire was conducted by Rear Admiral Forsyth Massey.
90	In 1967, investigations and reviews were conducted regarding safety in aircraft carrier operations. Reports were submitted and forwarded to various naval commands for analysis and implementation of recommendations.
91	The page highlights significant events related to the Russell Report recommendations, including a flight deck fire on the USS Enterprise and the establishment of the Carrier Aircraft Support Study (CASS) group. The report by Rear Admiral Bardshar validated the proposed damage control improvements, and updates on the recommendations were provided to the Chief of Naval Operations.

92   A machinery space fire on USS Saratoga resulted in casualties. The Chief of Naval Operations requested an update on the status of recommendations from the Russell Report, but periodic status reports were later discontinued.

93   This appendix summarizes several important recommendations for improving damage control on aircraft carriers, including advanced flight deck fire fighting systems, marking escape routes, improving interior communications, using improved fire hoses, reviewing ship alterations affecting safety, analyzing fires on specific carriers, and providing dedicated funding for damage control equipment.

94   The page contains a series of recommendations for improving safety measures on ships, including the requirement for multiple exits in berthing compartments, distributing information on gas mask capabilities, developing emergency breathing apparatus and better protective gear, providing training for damage control and fire fighting, and creating more realistic training aids.

95   This glossary provides definitions for various terms related to firefighting and ship safety on US Navy ships, including different classes of fire, watertight integrity conditions, and procedures for crew members. It also mentions the Manual of the Judge Advocate General Investigation and operating targets for purchasing items.

96   This page discusses various shipboard equipment and modifications, including oxygen breathing apparatus pilot landing aid television, PKP extinguishers, ship alterations, and William fittings

97   This page contains a bibliography of various sources related to naval accidents and safety measures.

98   This page lists various publications related to the investigation of ship emergencies, ship design, fire prevention, and memoirs.

99   This page lists various publications related to naval weapons, damage control, ship architecture, and safety in carrier operations.

100   The page includes references to a U.S. Army publication on military research and a book about the marine environment.

101   This page provides the initial distribution list for a document, including the addresses of various organizations and individuals involved in the distribution.

102   This page is a certification form for the distribution of a thesis on the impact of a fire on a US Navy ship. It includes information on the author, committee members, distribution statements, and limitation justifications.

103   This page provides information on different distribution statements that determine who is authorized to access and distribute certain documents. These statements are based on reasons such as protecting foreign information, proprietary information, critical technology, test and evaluation data, contractor performance evaluation, premature dissemination, administrative/operational use, software documentation, specific authority, and direct military support. The statements vary in their authorization scope, ranging from public release to limited distribution to specific government agencies or contractors.

THE IMPACT OF THE USS *FORRESTAL'S* 1967 FIRE ON UNITED STATES
NAVY SHIPBOARD DAMAGE CONTROL

A thesis presented to the Faculty of the U.S. Army
Command and General Staff College in partial
fulfillment of the requirements for the
degree

MASTER OF MILITARY ART AND SCIENCE
Military History

by

HENRY P. STEWART, LCDR, USN
B.S., Maine Maritime Academy, Castine, ME, 1992
M.S., Naval Postgraduate School, Monterey, CA, 1999

Fort Leavenworth, Kansas
2004

Approved for public release; distribution is unlimited.

MASTER OF MILITARY ART AND SCIENCE

THESIS APPROVAL PAGE

Name of Candidate: LCDR Henry P. Stewart, USN

Thesis Title: The Impact of the USS *Forrestal's* 1967 Fire on United States Navy Shipboard Damage Control

Approved by:

_____, Thesis Committee Chair
LTC Marian E. Vlasak, M.A.

_____, Member
CDR David Christie, M.M.A.S.

_____, Member
Jerold E. Brown, Ph.D.

Accepted this 18th day of June 2004 by:

_____, Director, Graduate Degree Programs
Robert F. Baumann, Ph.D.

The opinions and conclusions expressed herein are those of the student author and do not necessarily represent the views of the U.S. Army Command and General Staff College or any other governmental agency. (References to this study should include the foregoing statement.)

ABSTRACT

THE IMPACT OF THE USS *FORRESTAL'S* 1967 FIRE ON UNITED STATES NAVY SHIPBOARD DAMAGE CONTROL, by LCDR Henry P. Stewart, United States Navy, 112 pages.

This thesis examines the impact of the 1967 flight deck fire on the aircraft carrier USS *Forrestal (*CVA 59*)* and the resulting two investigations, on the development of US Navy damage control doctrine and equipment. The first investigation focused solely on the *Forrestal* fire; the second assessed the safety of aircraft carrier operations throughout the US Navy. Both investigation reports included several proposals to improve shipboard damage control. The thesis found that most of these recommendations were successfully implemented, substantially enhancing shipboard damage control capability over the long term. Successful implementation of these proposals depended on the following: strong support by, long-term involvement of, and resourcing by the Chief of Naval Operations, as well as broad agreement by senior Navy leaders that the proposed changes were necessary based on lessons learned from the two investigations. Training and material deficiencies appeared to be noncontroversial and thus relatively easy to correct; proposals that did not mesh well with Navy culture and existing personnel practices appeared especially controversial and were not successfully implemented.

ACKNOWLEDGMENTS

I would like to express my appreciation to the numerous people at the Naval Sea Systems Command and the Naval Operational Archives who helped provide invaluable data, particularly Mr. Douglas Barylski and Mr. Kenneth Johnson. I would also like to thank Lieutenant Colonel Vlasak, Commander Christie, and Dr. Brown for their guidance and advice throughout this project. I would like to thank Ms. Helen Davis for her help with proofreading the drafts of this thesis. Finally, I would like to thank my wife Amy for her patience and support.

TABLE OF CONTENTS

Page

MASTER OF MILITARY ART AND SCIENCE THESIS APPROVAL PAGE ii

ABSTRACT iii

ACKNOWLEDGMENTS iv

ACRONYMS vii

CHAPTER 1. INTRODUCTION 1

    The Ship 1
    The Incident 2
    Purpose of This Research 4
    Methodology 5
    Limitations 6

CHAPTER 2. DAMAGE CONTROL DOCTRINE 7

    World War II Damage Control Doctrine 8
    Postwar Doctrine Revision 11

CHAPTER 3. WARSHIP SURVIVABILITY FEATURES AND DAMAGE CONTROL GEAR 16

    Compartmentalizaton 16
    Enhanced Seagoing Capabilities 19
    Survivability Enhancements Based on World War II Experience 20
    Damage Control Equipment 22
    Personnel Protective Equipment 28

CHAPTER 4. RECOMMENDATIONS FROM INVESTIGATING PANELS 33

    The Judge Advocate General Investigation 34
    Findings of Fact 36
    Opinions and Recommendations 39
    The Russell Report 45
    Recommendations to Improve Warship Survivability Features 49
    Recommendations to Improve Personal Protective and Damage
      Control Equipment 53
    Recommendations to Increase Damage Control Awareness and Training 57

CHAPTER 5. IMPLEMENTATION OF REPORT RECOMMENDATIONS 63

    Implementation of *Forrestal Fire Investigative Report's* Recommendations 63
    Implementation of the *Russell Report's* Recommendations 68
    Impact of the *Enterprise* Fire on Russell Panel Recommendations 70

CHAPTER 6. CONCLUSION 82

    Lasting Impact of the *Forrestal* Fire..82
    Lasting Impact on Doctrine 83
    Lasting Material Impact..84
    Unsuccessful Damage Control Improvement Ideas..85
    Implications for Today's Navy..87

APPENDIX A. TYPICAL NAVY FIREMAIN "LOOP" DIAGRAM............................90

APPENDIX B. CHRONOLOGY OF EVENTS..91

APPENDIX C. SUMMARY OF SELECTED *RUSSELL REPORT*
RECOMMENDATIONS 95

GLOSSARY 97

BIBLIOGRAPHY...99

INITIAL DISTRIBUTION LIST 103

CERTIFICATION FOR MMAS DISTRIBUTION STATEMENT 104

ACRONYMS

CASS	Carrier Aircraft Support Study
CCOL	Compartment Check-Off List
COMNAVAIRLANT	Commander, Naval Air Force, US Atlantic Fleet
COMNAVAIRPAC	Commander, Naval Air Force, US Pacific Fleet
CVA	Carrier Vessel Attack
JAGMAN	Manual of the Judge Advocate General
NSTM	Naval Ship's Technical Manual
OBA	Oxygen Breathing Apparatus
OPTAR	Operating Target
PKP	Potassium Bicarbonate
PLAT	Pilot Landing Aid Television
PRSACO	Panel to Review Safety in Aircraft Carrier Operations
SHIPALT	Ship Alteration

CHAPTER 1

INTRODUCTION

One of the most serious disasters in modern naval history began just before 11:00 a.m. on 29 July 1967. On that morning, one of the United States Navy's most modern aircraft carriers, USS *Forrestal* (CVA 59) was operating in waters off the coast of Vietnam.

The Ship

Forrestal was the first of the "supercarriers" of the US Navy. Commissioned in 1955, she was the first US aircraft carrier specifically designed to operate jet aircraft, and was the first carrier the United States built following World War II. Her namesake was James V. Forrestal, a former naval aviator, and our nation's first Secretary of Defense. *Forrestal* was 1,076 feet long, 252 feet wide at her flight deck, and displaced over 79,000 tons. In comparison, the *Essex* class aircraft carriers built during World War Two only displaced 41,000 tons. *Forrestal*'s flight deck had approximately 250,000 square feet of area. Her engineering plant was able to produce 260,000 horsepower and consisted of oil-fired boilers and steam turbines. She had four propellers, and could achieve a top speed of greater than 30 knots (approximately 35 miles per hour). She had 19 separate levels (called "decks" in naval terminology), and over 2,000 separate compartments, or "spaces." A crew of 3,000 men operated the ship, and 2,500 more men operated and maintained the embarked aircraft. *Forrestal* had her own post office, laundry rooms, and ship's store (selling cigarettes, snacks, and personal items for crew members), staterooms for officers, and lounges for the crew. She produced her own electricity and distilled approximately 200,000 gallons of fresh water daily for drinking, washing, and cooking. Many of her interior compartments were air-conditioned. She was a virtual "city at sea."

The Incident

The *Forrestal* had recently arrived in the waters off Vietnam, and had been bombing targets in North Vietnam for the previous four days. *Forrestal* launched and recovered all aircraft from the first strike of the day without incident, and the crew prepared the second strike group's aircraft for launch. Crewmen staged 27 aircraft on the flight deck. The fully armed planes were crowded together on deck as the crew conducted final preflight checks. Each aircraft carried a full load of bombs, rockets, and ammunition, and the fuel tanks of each plane were full. In addition, crew members staged several tons of bombs on the flight deck on wooden pallets.

The *Forrestal* accelerated to nearly 30 knots and turned into the wind as she prepared to launch the second strike of the day (she was generating high relative winds over her flight deck to provide sufficient lift to safely launch her aircraft.) Several of the jets started their engines in preparation for launching. Without warning, a rocket was accidentally fired from one of the F-4 Phantom fighter planes on the deck. The rocket struck a crewmember on deck before striking and ripping open an A-4 Skyhawk staged on the opposite side of the flight deck. The rocket passed through the aircraft without exploding and hit the ocean. However, several hundred gallons of jet fuel poured from the Skyhawk's punctured fuel tank and quickly ignited by particles of burning rocket propellant left on the flight deck. The burning fuel from the stricken jet was pushed aft (back) by the heavy winds across the flight deck. The burning fuel quickly engulfed several other aircraft staged on the flight deck. Within seconds, these aircraft began burning, and the fire continued to spread. The officer of the deck (the officer on watch responsible to the commanding officer for safe operation of the ship) immediately sounded General Quarters. This was a shipwide announcement that the ship was experiencing an emergency. He quickly followed this up with a verbal report over the 1MC (the shipwide general announcing system) notifying the crew of the fire on the flight deck. The *Forrestal's* crew moved toward their assigned "battle stations."

When General Quarters was set, *Forrestal's* crew members fully manned all positions in the ship's damage control organization. The crew also set Material Condition Zebra. This compartmentalized the ship by closing doors and hatches throughout the ship. Many of these hatches were normally open to facilitate crew movement throughout the ship. Closing them would help to limit the flow of smoke, fire, and firefighting water through the ship. The Commanding Officer ordered the ship to stop, to reduce the wind across the flight deck that was fanning the blaze. However, the fire continued to spread quickly.

The heat of the fire exploded a bomb on the flight deck approximately ninety seconds after the fire began, and a second bomb exploded a few seconds later. These explosions severely damaged the carrier and killed several sailors on the flight deck. The fuel tanks of several other planes ruptured, adding to the intensity of the blaze. The exploding bombs created several holes in the flight deck, allowing fire and smoke to spread into the interior of the ship.

Forrestal's crew feverishly battled and eventually extinguished the fire. It took over twenty-four hours to extinguish the fires that spread below the flight deck. The losses caused by this incident were high. One hundred thirty-four sailors were killed by the fire, and 161 more were injured. Over twenty aircraft were lost. The damage forced *Forrestal* to suspend combat operations and conduct temporary repairs in the Philippines before returning to the US for permanent repair. Repairs to the ship cost approximately $72 million, and took approximately two years to complete.

Purpose of This Research

Sailors have feared fires at sea since the days of the earliest ships. Even in modern times, ship's crews had to depend on each other to save their ship (and their own lives) when disaster struck. Every sailor had to be a firefighter as well. Proficiency was important, since fire could quickly spread in the hazardous shipboard environment. It remains vital for the Navy to accurately assess the cause of disasters and apply lessons learned to prevent similar situations from recurring. Failure to do so can result in many lives lost, millions of dollars in damages, and even the loss of a ship.

The purpose of this thesis is to examine how the Navy applied lessons learned from the USS *Forrestal* conflagration on 29 July 1967 to improve fleetwide damage control capability (training, doctrine, installed equipment, and warship design). The primary research question is: Did this fire have significant influence on the US Navy's damage control doctrine and training, shipboard firefighting equipment, and warship construction? Secondary research questions include: If so, what specific changes resulted from this disaster? Were these changes significant and permanent? Does historical evidence show that these changes were effective?

The thesis statement of this research is that the US Navy significantly improved damage control training, damage control equipment, and warship design as a direct result of lessons learned from the fire on USS *Forrestal*.

Methodology

Numerous historical documents were examined to prove the thesis and answer the research questions. The official investigation report of the incident was studied to learn about the damage control organization on *Forrestal* in 1967. This thesis reviewed what damage control equipment was available to *Forrestal's* crew, what survivability features were included in *Forrestal* by designers, and what damage control doctrine existed to guide her crew. The thesis also examined the issue of whether the Navy built damage control improvements into *Forrestal* because of lessons learned from previous disasters or battle damage.

Two official Navy investigation panels were convened as a direct result of the fire on *Forrestal*. This thesis reviewed the recommended changes to improve damage control on US Navy ships submitted by these panels. The following specific areas were examined: What specific changes did the panels recommend? Were they implemented? Were these changes effective? Did shipbuilders apply lessons learned from the *Forrestal* fire to incorporate design changes into future warships? If so, what changes did they make, and how did these changes improve a ship's damage control capability? Did the Navy only apply design changes to ships built after the *Forrestal* fire, or did they make some changes to improve the damage control capability of existing ships?

The thesis examined a similar fire that occurred on the US Navy aircraft carrier *Enterprise* in 1969 (approximately eighteen months after the *Forrestal* fire) to assist in assessing whether lessons learned from the *Forrestal* were significant and enduring.

This thesis answered the following questions. Did the damage control organizations of these ships benefit from lessons learned from *Forrestal's* fire? Did shipbuilders incorporate improved damage control features into these ships? If so, did these changes serve to mitigate the effects of damage?

Limitations

There were numerous limitations to this research. The study focused on the specific lessons learned from the fire on USS *Forrestal* on 29 July 1967, and how these lessons were applied by the US Navy to improve damage control capabilities on its warships in later years. Major themes of interest included examining what damage control doctrinal, shipboard firefighting and damage control systems changed from analysis of this fire. This research also discussed design changes the Navy made to its warships after analyzing this disaster. This study briefly examined incidents that occurred on US Navy warships after the *Forrestal* incident to determine if the US Navy successfully applied these lessons. This research did not study damage control doctrine, equipment, or ship design in foreign navies. The study was limited to the impact that this incident had on damage control on US Navy surface warships. This study relied on official Navy accounts of the fire, reports of official Navy panels convened to review the fire, and Navy damage control doctrine and instructions.

CHAPTER 2

DAMAGE CONTROL DOCTRINE

Firefighting and damage control have been important to the US Navy since the age of sail. This concern remained vitally important in 1967, since naval ships contained large quantities of fuel, oils, weapons, ammunition, paint, and many other hazardous and flammable materials. Other factors also elevated the risk of fire and damage--ships launched and recovered helicopters and other aircraft, frequently maneuvered at high speeds in close proximity to other vessels, and steamed in widely variable weather and sea conditions. The danger to ships from accidental fires and flooding was high whether the ship was operating in home waters or was forward deployed to war zones. Fire or other damage usually struck suddenly, and had to be quickly controlled to prevent extensive damage to the vessel and minimize injuries to her crew. Perhaps the most important aspect of damage control was that any ship sustaining damage often had to rely completely on its own crew to take responsive action. Operational circumstances demanded that naval vessels often operated independently of other ships, and weather could prevent other ships from assisting.

Well-known naval authorities and retired US Navy Captains John V. Noel and Edward L. Beach provide one authoritative definition of damage control. Captain Noel revised *Knight's Modern Seamanship, The Division Officer's Guide, The Watch Officer's Guide, Ship Handling*, and coauthored *Naval Terms Dictionary* with Captain Beach. Captain Noel commanded a destroyer, supply vessel, and cruiser during his long career. Captain Beach served as a damage control assistant and chief engineer in submarines, and also commanded several submarines. He wrote several fictional and nonfiction works, and was well known for his novel *Run Silent, Run Deep*. Captains Noel and Beach define damage control as "Measures necessary to preserve and reestablish shipboard watertight integrity, stability, maneuverability and offensive power; to control list and trim; to make rapid repairs of materiel, to limit the spread of and provide adequate protection from fire; to limit the spread of, remove the contamination by, and provide adequate protection from toxic agents; and to provide for care of wounded personnel."[1] Although many of the procedures used to combat damage control changed substantially over time, the basic problems remained constant. The US Navy relied on training (damage control schools, shipboard drills) and doctrine (official publications promulgating techniques and procedures to be used in controlling damage). Doctrine evolved over the years to reflect advances in damage control equipment technology, changes in ship design, and to incorporate lessons learned from earlier incidents.

World War II Damage Control Doctrine

US Navy damage control doctrine in effect during the 1967 *Forrestal* fire evolved from the Navy's World War Two era damage control doctrine. The American Navy's primary shipboard firefighting doctrine during the Second World War was the *Fire-Fighting Manual (Naval Ships Publication 688)*. This 133-page manual was published in 1943 to provide a sound basis for naval firefighting and damage control to the many inexperienced personnel joining the rapidly expanding wartime navy. It described the nature and hazards of the shipboard environment, explained how to use the Navy's shipboard damage control and personnel protective equipment, and detailed the techniques and procedures necessary to fight fires and control damage. Although the *Fire-Fighting Manual* was useful in familiarizing Navy officers and enlisted men with the equipment, techniques, and procedures necessary to combat fires and damage on ships, the Navy also operated seven major shipboard Fire Fighter's Schools on the larger naval bases. The Navy's Bureau of Ships developed and prescribed the course of instruction taught at these schools to standardize the training. Course lengths of one to ten days were available. The full (ten-day) course included instruction on the various types of fires likely to be encountered on ships, training on all Navy damage control equipment (instructor would demonstrate how to use each item, and students would then practice using it), and extinguishment of actual fires and repair of simulated damage in simulated ship compartments. The shorter courses focused on familiarization and practice with shipboard damage control equipment. In 1943, approximately 600 students per month were attending each of the seven Navy Fire Fighter's Schools.[2]

Shortly before the war ended, the Bureau of Ships published a *Handbook of Damage Control* that detailed many of the damage control lessons that had been learned by the Navy during the war years. The first nine pages of this manual were exclusively composed of excerpts from US Navy war damage reports. These excerpts provided examples of a warship's inherent resistance to damage, the importance of maintaining watertight integrity, particularly effective fire prevention measures and firefighting actions taken by the crews of several warships, and the importance of damage control training and personnel protection.[3]

In addition to the *Handbook of Damage Control*, the Navy's Bureau of Ships compiled several reports in the mid-to-late 1940s analyzing the damage incurred on US Navy ships during the Second World War. These reports were based on accounts of shipboard personnel, reports of observers stationed on other ships, and assessments of damage conducted by Bureau of Ships and shipyard personnel when damaged ships returned to port. Each volume in this series of damage reports was dedicated to a particular type of ship, such as destroyers, cruisers, battleships, or aircraft carriers. These reports described the types of damage sustained by each ship, what weapons caused the damage, what structural and hull damage was sustained, how buoyancy and stability were affected and what fires and flooding resulted and analyzed the performance of the crews in controlling the damage. These reports also detail some of the improvements in damage control procedures and equipment developed as a result of wartime experience.[4]

The report on destroyers is particularly illuminating because destroyers were the most numerous type of combatant vessel in the US Navy during the Second World War (377 were in commission in 1945). The *Destroyer War Damage Report* stated that the Navy suffered severe losses due to fires during the first year of the war. The report also stated that firefighting performance improved throughout the war as a result of several factors. First, avoidable fire hazards (excess flammable materials) were removed from Navy ships. Second, ships were given an increased allowance of firefighting equipment. This new equipment tended to be more effective than the old equipment, and was widely dispersed throughout the ship to increase rapid accessibility when needed. Third, damage control lessons learned were reinforced in the Navy's firefighting schools. Finally, the *Destroyer Report* concluded that:

> In general, the firefighting performance of destroyer crews in the latter part of the war, utilizing their improved training and newly developed equipment, was very encouraging. Their record proved that speed in getting water to the fire is all-important and is the mark of effective drilling. One hose stream brought to the scene of the fire within a minute often proved more valuable than several a few minutes later. Drills in immediately running hose and rigging portable pumps

for use in the damage area and in promptly checking the intactness of the firemain repeatedly proved their value.[5]

Postwar Doctrine Revision

The next major revision of US Navy firefighting doctrine was issued in May 1951, when *Bureau of Ships Manual Chapter Ninety-Three: Fire Fighting – Ship* was published. This new manual replaced the old *Fire-Fighting Manual*, which was last revised in 1944. The new manual reflected more changes in equipment and procedures made as a result of lessons learned from the Second World War. *Chapter Ninety-Three* consisted of 113 pages, broken down into three sections. The first section discussed the firefighting and damage control equipment available to the shipboard firefighter. The second section described how to properly use shipboard personnel protective equipment, and the final section prescribed firefighting techniques and procedures.[5]

The next version of the Navy's firefighting manual, *Naval Ships Technical Manual Chapter 9930: Fire Fighting – Ship* (referred to hereafter as *NSTM 9930*) was issued approximately one month after the *Forrestal* fire, on 1 September 1967. Although it was not in effect during the *Forrestal* fire, it does illustrate the state of development of Navy Damage Control doctrine at the time of the incident (It didn't include any lessons learned from the *Forrestal's* fire, since that incident was still under active investigation). This initial version of *NSTM 9930* contained the same three sections as *Bureau of Ships Chapter 93*, but Section Two (Protective Equipment) was a placeholder, with no information included. The overall document was reduced to ninety-nine pages. The first seventy-three pages were dedicated to the nature of fire and firefighting equipment; the remainder dealt with firefighting techniques and procedures.[7] Significantly, most of the material describing fire, firefighting agents and shipboard firefighting equipment included in *NSTM 9930* was virtually identical to discussions in the older doctrine. Although warships had dramatically increased in size and complexity since World War Two, it seemed that the damage control tools available to sailors had not significantly changed.

It is important to note that the 1967 version of *NSTM 9930* was not designed as a stand-alone reference document for shipboard firefighting. For the first time, the 1967 edition of *NSTM 9930* directed ships to establish and maintain a reference library of damage control publications, and contained a list of forty-six separate publications to be included in this library. This list included a Ship's Damage Control Book (tailored to each type of Navy ship in service), a complete set *of Naval Ship's Technical Manuals* (each volume, or chapter, provided information on a particular aspect of Navy operations), instruction manuals on damage control and personnel protective equipment used aboard naval vessels, and naval regulations and instructions governing damage control.[8] Of course, the usefulness of this reference library depended largely on how effectively each ship's senior damage control experts integrated the material into their damage control training program.

NSTM 9930 stressed the importance of reducing fire hazards to decrease the risk of shipboard fires and to minimize the damage sustained when a fire did occur. It prescribed four basic principles to reduce unnecessary fire hazards: first, proper stowage of combustible materials; second, regular and frequent inspections of shipboard spaces by shipboard leaders; third, training all personnel on the importance of reducing fire hazards; and finally, strict enforcement of fire prevention policies and practices.[9]

NSTM 9930 also placed heavy emphasis on the importance of frequent, realistic training drills to improve the efficiency of a ship's damage control organization:

> Every man in the organization must know where to go, how to get there, what may be needed, and what to do upon arriving at the scene of a fire. It is only by constant drilling that fire-fighting parties can learn to function as teams. Men must be trained to act immediately and use the proper equipment and correct procedure. . . . Drills uncover weaknesses and failures of personnel and material which can be eliminated or recognized as a possible source of danger should an actual fire occur in the area. . . . An effective protection against fires in ships in the quantity and quality of training before a fire starts.[10]

The third section of the 1967 *NSTM 9930, Fire Fighting and Fire Hazards*, was significantly different than earlier doctrine. In the older doctrine, this section discussed the nature of shipboard fires and the effectiveness of extinguishing agents, such as solid-stream water, water fog, foam, carbon dioxide, and others. After this discussion, the doctrine stipulated appropriate techniques and procedures to combat several common types of shipboard fires (such as flight deck fires, engine room fires, and fires in electronic equipment rooms). The 1967 *NSTM 9930* contained this information as well, but it also included an entirely new subsection on the configuration of the ship's damage control organization. It directed each ship to implement tailored "Fire Bills." Fire Bills were published lists that assigned specific duties and responsibility to specific crew members in the event of a fire. Rudimentary fire bills had been in use since the Age of Sail, but the increased size and complexity of modern warships demanded a highly specialized list. Examples of positions on a typical fire bill include nozzlemen (responsible for manning the nozzle end of the hose and attacking the fire), hosemen (who maneuvered the hose to support the nozzleman), plugmen (who opened valves charging the hoses), investigators (who rapidly surveyed the ship to determine the location and extent of damage), and scene leaders (who directed local damage control efforts and reported status of those efforts up the chain of command). Crew members received training to qualify for positions on the fire bill. Sailors were required to qualify for these positions sequentially. For example, a newly reported sailor could quickly qualify as a plugman. As a plugman, this junior sailor would only be responsible for operating a valve feeding a single fire hose. With more experience, the plugman would qualify to serve as a hoseman, then as a nozzleman. A scene leader was required to be proficient in all of these junior positions. Separate Fire Bills were required for periods when the ship was at sea and when the ship was inport. The entire ship's company was available to participate in damage control efforts while the ship was underway, but a much smaller number of personnel were available inport. While the ship was inport, the majority of crew members departed the ship after normal working hours. The ship's company was split into several "duty sections." Each duty section would spend the night aboard to oversee the ship until relieved by the next duty section the following day. These duty sections were comprised of relatively small portions of the overall ship's company, and would only man a single repair locker to respond to emergencies (all repair lockers were manned if required during emergencies at sea). The duty section would frequently be augmented during fires inport (many sailors lived aboard ship), but the fire bill provided supervisory personnel with a formal list of qualified sailors charged with responding to damage occurring during their duty day. The engineer officer (officer in charge of the Engineering Department, and the individual who, by Navy Regulations, was also designated as the damage control officer) was responsible for supervising the Fire Bills and ensuring that assigned personnel were properly trained and qualified for their positions.[11] *NSTM 9930* also provided several examples of typical shipboard damage control organizations, defining required positions and responsibilities of assigned personnel and delineating necessary types and quantities of damage control equipment.

The next chapter examines the survivability features that were included in US Navy warships in general and the *Forrestal* in particular as a result of experience and lessons learned from previous incidents and battle damage. The chapter also describes the damage control equipment and personnel protective gear used by shipboard firefighters in 1967.

---

[1] John V. Noel and Edward L. Beach, *Naval Terms Dictionary* (Annapolis, MD: UNITED STATES Naval Institute, 1971), 83.

[2] Navy Department, *Fire-Fighting Manual: NAVSHIPS PUB 688* (Washington, D.C., Bureau of Ships, 1943), 132.

[3] Navy Department, *Handbook of Damage Control: NAVPERS PUB 16191* (Washington, D.C., Bureau of Ships, 1945), 1-9.

[4] Navy Department, *War Damage Report No. 51, Destroyer Report: Gunfire, Bomb, and Kamikaze Damage, Including Losses in Action, 17 October, 1941 to 15 August, 1945* (Washington, D.C., Bureau of Ships, 1947), 1.

[5] *War Damage Report No. 51*, 17-19.

[6] Navy Department, *Bureau of Ships Manual Chapter 93: Fire Fighting – Ship* (Washington, D.C., Bureau of Ships, 1951), 1.

[7] Naval Ship Systems Command, *Naval Ships Technical Manual Chapter 9930: Fire Fighting – Ship* (Washington, D.C., Naval Ship Systems Command, 1967), 1.

[8] Ibid., 1-2.

[9] Ibid., 1.

[10] Ibid., 2-3.

[11] Ibid., 75.

CHAPTER 3

WARSHIP SURVIVABILITY FEATURES AND DAMAGE CONTROL GEAR

Survivability was one of the warship's primary design considerations. Warships were designed to survive and operate effectively in extremely inhospitable conditions at sea. Heavy seas exerted tremendous stress on a ship's structure, and were often encountered with little warning. In February 1933, the USS *Ramapo* survived an encounter with a 112 feet high wave in the Pacific Ocean (the highest ever reliably reported, according to Professor Jerome Williams, who published several works on oceanography and originated the oceanography course at the US Naval Academy).[1] Although this is an extreme example, it illustrates the harshness of the marine environment even in the absence of accidental fires or enemy action. All ships that are expected to perform well in these demanding conditions require a high degree of buoyancy and stability. However, naval vessels must be built stoutly enough to sustain damage and remain operational, so they require even greater protection than would normally be expected. The elements of survivability considered by naval architects that designed warships such as the *Forrestal* included compartmentalization, seagoing capability, and improvements based on experience gained during the Second World War.

Compartmentalizaton

Shipbuilders have always been concerned with the hazards of flooding and sinking. Even wooden ships would easily sink if their interior compartments were flooded. This concern intensified as ships were built with steel hulls, and their size increased dramatically. Disasters such as the loss of the *Titanic* emphasized the importance of compartmentalization, or subdividing a ship's structure into numerous watertight compartments.

Warships required an inherent ability to resist damage caused by underwater attack (such as damage from naval mines or torpedoes). Transverse watertight bulkheads (connecting the port and starboard sides of the hull) are effective in containing flooding along the length of a ship's hull after underwater damage is sustained. By the time *Forrestal* was built, all warships contained a series of numerous transverse bulkheads extending from the keel (bottom) of the ship to the main deck (frequently termed the damage control deck). The forward most transverse bulkhead was generally placed several feet abaft (behind) the bow. It was specifically designed to reduce a ship's vulnerability to flooding as a result of collisions, and was termed the collision bulkhead.[2] The exact location of the collision bulkhead varied widely depending on the ship's length. Designers termed the imaginary vertical line extending through the point where the ship's bow met the sea the "forward perpendicular." Similarly, the vertical line extending through the point where the stern touched the water was termed the "after perpendicular." The length between these two imaginary lines was referred to as the "length between perpendiculars," and the collision bulkhead was located at least 5 percent of this length abaft the forward perpendicular. Longitudinal watertight bulkheads ran fore and aft between main transverse bulkheads. Longitudinal bulkheads were often used to protect vital spaces (containing equipment essential to operate the ship) from flooding. Longitudinal bulkheads had to be carefully designed to minimize unsymmetrical spaces in the ship's hull. Unsymmetrical spaces resulted when the compartmentalized spaces on one side of the ship's centerline were not identical in volume to those on the other side. The ship's stability decreased if an unsymmetrical space flooded.

In addition to limiting progressive flooding (the spread of flooding throughout the ship), compartmentalization was useful in limiting the spread of fire and smoke through the ship's interior spaces. The Navy developed several procedures and requirements designed to maximize the effectiveness of compartmentalization. Many compartments had necessary fittings, such as doors, hatches, ventilation ducts, and electrical cables that passed through watertight bulkheads. Regular inspection and maintenance was required to ensure that these fittings did not reduce a ship's watertight integrity. Compartment Check-Off Lists (CCOLs) were developed, listing each of these fittings in every compartment. Regular inspections of items listed on the CCOLs were required, and periodic maintenance was required on items susceptible to wear, such as door gaskets.[3]

The US Navy also developed three major material conditions of readiness for all vessels. Each material condition provided a different degree of tightness and protection. Crew members labeled all fittings (sometimes referred to as closures) to facilitate rapid identification. Condition "X-Ray" allowed the most fittings, such as doors, hatches, and scuttles, to remain open. This increased the convenience and ease with which personnel could transit throughout the ship, but also provided the least degree of protection against the spread of fire, smoke, or flooding. Condition "X-Ray" was normally set inport during normal working hours when the ship was not believed to be at risk from attack. Condition "Yoke" required more fittings to be closed, and consequently provided more protection. Condition "Yoke" was typically set at all times while the ship was at sea and after normal working hours in port. Condition "Zebra" provided the most protection, and required most fittings to be closed. Condition "Zebra" was normally set when the ship expected to enter combat soon (General Quarters was set), or in the event of fire and flooding in the vessel. Condition "Zebra" was not normally set for long periods at sea, since it significantly hampered the movement of crew and material throughout the ship, and reduced crew comfort since most ventilation was secured during Condition "Zebra." Modifications of these three basic conditions, such as "Circle X-Ray, Yoke, and Zebra" permitted certain predesignated closures to be opened by crew members. This allowed crew members to transit through zones, and facilitated moving ammunition and other supplies throughout the ship. "William" fittings were essential to the ship's mobility and fire protection. These fittings were marked with a black "W," and were kept open during all material conditions. Fire pump and other vital pump cutout valves were classified as "William" fittings.[4]

Enhanced Seagoing Capabilities

When she was commissioned in 1955, *Forrestal* was the world's largest aircraft carrier. Her large size greatly enhanced *Forrestal's* seagoing capabilities, since a warship's inherent survivability and seaworthiness tend to increase with the vessel's size. For example, a larger ship generally has more watertight compartments than a smaller ship. Reserve buoyancy, the volume of the watertight portion of the ship above the waterline, is also usually greater for larger ships.[5] As a result, larger ships are inherently able to sustain more damage and remain afloat. Larger ships also enjoy several other characteristics useful in naval vessels. A smaller fraction of the ship's displacement is required for propulsion equipment and fuel storage on larger ships (or a greatly extended range is possible if the same percentage of fuel to ship's displacement is maintained), and larger ships generally are capable of higher speeds in rough seas.[6] Larger ships are also capable of carrying more weapons, equipment, and stores. Naval vessels were limited in size by treaties for much of the interwar period, but began to increase in size in the late 1930s. This trend toward increasing warship size was still continuing when *Forrestal* was built in the early 1950s. The Forrestal displaced 79,000 tons and contained 1,240 watertight compartments; while the Essex Class carriers built during World War Two displaced less than 40,000 tons and contained 750 watertight compartments.[7] The trend toward increasing warship size was not limited to aircraft carriers – many combatant ships in the US Navy were increasing with size during this period. For example, the *Porter* class destroyers of the 1930s displaced approximately 1,850 tons, the *Fletcher* class destroyers of the 1940s displaced over 2,500 tons, and the early 1960s *Charles F. Adams* class of destroyers displaced nearly 3,400 tons.[8]

<u>Survivability Enhancements Based on World War II Experience</u>

Several survivability features recommended by the Navy's World War Two damage reports were incorporated in *Forrestal*. *Forrestal* was built with an armored flight deck, constructed of thick, high-strength steel. World War Two experiences showed that this would decrease the amount of structural damage sustained in interior compartments from explosions or fires on the flight deck.[9]

Forrestal was also equipped with a firemain loop. The firemain loop was designed to correct a serious deficiency observed during the Second World War, when many crews were unable to combat shipboard fires because firemain pressure was lost as a result of damaged piping. In several instances fire pumps continued to run and the ship's stability was reduced by tons of seawater flowing into interior compartments from damaged piping.[10] A firemain loop was a line of saltwater piping that ran continuously around the vessel. The loop also incorporated several runs of piping running athwartships (connecting the firemain piping on the ship's port side with that on the starboard side). These transverse piping runs were placed near the bow, amidships (near the center of the vessel), and aft. The loop could be charged with several fire pumps, located in numerous compartments throughout the ship. Cutout valves were placed at regular intervals in the piping runs. This arrangement enabled the ship's crew to isolate damaged portions of the firemain, while still supplying firefighting water where needed. The dispersion of multiple firefighting pumps helped to ensure that adequate firemain pressure could be maintained even if some pumps were damaged or inoperable. If the ship expected to enter combat, several isolation valves would be closed near the transverse piping runs to create several smaller firemain loops. This would ensure firemain pressure to most of the ship in the event of firemain piping damage, and would limit the amount of flooding sustained from broken piping. A diagram of a typical firemain loop is included in Appendix A.

Flight deck and hangar deck sprinkler systems were also installed on *Forrestal* to cool ordnance during fires (to prevent cook-off) and to help prevent the spread of fires in these areas. Several high capacity foam-generating stations were also installed. These stations were capable of generating large amounts of firefighting foam to help smother fires in the hangar deck or on the flight deck. US Navy damage reports from the Second World War indicated that all of these features proved to be effective in limiting damage during actual fires.[11]

Forrestal was also designed to carry aircraft using JP-5 for fuel. JP-5 was much less volatile than the aviation gasoline that had been carried aboard aircraft carriers in World War Two, and was considered to be less hazardous for shipboard use.

As the last several pages have shown, naval warships such as *Forrestal* were designed to sustain damage and survive. However, another significant component of damage control was found in the development of an extensive array of specialized equipment. This equipment ranged from items designed to be operated by individual crew members, to larger systems operated by a team. Some of this equipment was used to control and extinguish fires, combat flooding, and isolate damaged systems. Personnel protective equipment helped reduce the risk to crew members as they fought to control damage in hazardous environments. The next two sections of this chapter will examine the damage control and personnel protective equipment available to *Forrestal's* crew.

Damage Control Equipment

The equipment shipboard firefighters used to extinguish fires depended largely on the class, or type, of fire. Class Alpha fires involved combustible materials such as bedding, books, and clothing. Class Alpha fires left embers, which made these fires highly susceptible to rekindling. Water was the firefighting agent of choice for Class Alpha fires, since it lowered the temperature of the burning items and helped prevent reflashes.

Class Bravo fires involved burning flammable liquids, such as fuel oils, paint, and lubricants. They did not leave embers, and could be effectively extinguished by using firefighting foam to create a barrier between the burning liquid and the air needed for continued combustion.

Class Charlie fires occurred in electrical equipment. Carbon dioxide was the agent of choice for Class Charlie fires for two primary reasons: it would not damage the equipment, and it reduced the hazard of electrical shock for firefighters.

Class Delta fires occurred when metals such as magnesium ignited. NSTM 9930 stated that no effective firefighting agents existed for Class Delta fires. Burning metals were generally jettisoned if possible.[12]

To combat this array of possible conflagrations, shipboard firefighters had an extensive amount of available equipment. The fire main delivered firefighting water to fireplugs and sprinkler systems throughout the ship. Most fireplugs on aircraft carriers had outlets 2 ½ inches in diameter. Some plugs had 1 ½ inch reducing connections installed. These reducing connections would either have a single outlet, or would use a double Y-gate connection with two 1 ½-inch outlets. The fireplugs on *Forrestal* were positioned so that any point on the ship could be reached with a one hundred-foot length of hose from at least two separate locations. One hundred feet of hose was always connected to each fireplug. Specialized wrenches, termed spanners, were placed near each fireplug to connect additional hose sections as needed.

A Navy all-purpose nozzle was attached to the end of each hose connected to the ship's firemain. All-purpose nozzles could deliver either solid streams of firefighting water, or fog. Four, ten, and twelve-foot long applicators could be inserted into the end of an all-purpose nozzle to provide low-velocity fog. Solid streams of firefighting water were effective against Class Alpha fires, while water fog was useful against both Class Alpha and Bravo fires. Water fog was also used to help shield personnel from the heat of shipboard fires, and to cool munitions to prevent cook-off.[13]

Firefighting foam was very useful in fighting Class Bravo fires. In 1967, two basic types of foam were available in the US Navy. One type was termed protein foam since it consisted of a hydrolyzed protein base; the other type was called "Light Water," and was composed of a mixture of fluorinated surfactants. Both types came in concentrated liquid form, and six parts of concentrate were mixed with ninety-four parts of water to create firefighting foam. The two types of foam were fully compatible, but the Navy planned to gradually phase out the protein foam since it had a limited shelf life. The Light Water concentrate could be stored indefinitely before use.[14]

Naval vessels had several means of generating and delivering firefighting foam. The simplest piece of equipment used was a mechanical-foam nozzle with a pickup tube. A firefighting hose was connected to the nozzle, and the pickup tube was inserted into a five-gallon foam concentrate container. When the hose was charged, water flowing through the nozzle would create suction, drawing the concentrate up into a mixing chamber in the nozzle. The mixing chamber was sized to mix air, water, and foam concentrate together in the proper proportions to create firefighting foam. The mechanical-foam nozzle would empty a five-gallon foam container in about ninety seconds, producing approximately 660 gallons of foam in that time. Additional concentrate cans could be placed nearby if more foam was required.[15]

Larger pieces of equipment, known as proportioners, were used to protect machinery spaces, aircraft hangars, and flight decks. Proportioners used water motors and liquid foam pumps to generate foam. The size of the motors and pumps were designed to maintain the necessary proportion of foam concentrate to water. These proportioners consisted of dedicated firemain piping to supply water, fixed foam concentrate tanks, and supplied foam to hose stations as well as sprinkler heads. Although the larger foam stations could be started remotely, a crew of three or four sailors was assigned to monitor and operate each station. This crew would establish communications with the hose station near the fire, and would replenish the foam tank with additional concentrate as needed. The size of the tank varied by station--smaller proportioners had fifty-gallon liquid concentrate tanks, while the larger stations had 300-gallon tanks. The high-capacity foam stations serving the hangar and flight decks could produce 5,700 gallons per minute of foam at maximum output. The 300-gallon foam concentrate tank would be emptied in just over five minutes at this rate. Sailors would have to continuously empty five-gallon cans of concentrate into the liquid foam tank (at the rate of fifty-seven gallons per minute) to keep each high-capacity foam station operating.[16]

Two common types of portable fire extinguishers were also carried aboard Navy ships. These extinguishers used carbon dioxide or dry chemicals as extinguishing agents, and were placed at frequent intervals along the bulkheads of passageways and in many compartments throughout naval vessels.

Standard navy portable carbon dioxide extinguishers contained fifteen pounds of pressurized agent. They were effective against small Class Alpha, Bravo, or Charlie fires, had an effective range of three to five feet, and lasted forty to forty-five seconds. The carbon dioxide provided very little cooling effect, so larger fires were very susceptible to reflash after being extinguished. However, their small size and ubiquity throughout naval vessels allowed crew members to rapidly deploy them against small fires before the ship's damage control organization could respond with more substantial equipment.[17]

Portable dry chemical extinguishers, known as PKP extinguishers, were also installed in large numbers throughout naval vessels. These extinguishers used a small carbon dioxide charge to expel eighteen pounds of potassium bicarbonate based agent. The dry chemical extinguishers had an effective range of eighteen to twenty feet, and would last from eighteen to twenty seconds. These extinguishers were primarily intended for use against small Class Bravo fires, but could also be used to extinguish Class Charlie fires. The dry chemical agent was approximately four times more effective than an equal weight of carbon dioxide against flammable liquid fires, but left a fouling residue on electrical equipment when used on Class Charlie fires. Like carbon dioxide extinguishers, the dry chemical agent provided very little protection against reflash. It was intended only to extinguish small fires, or to help extinguish larger fires in conjunction with firefighting foam.[18]

The *Forrestal* was also equipped with emergency pumping equipment, intended to augment or temporarily replace damaged portions of the ship's firemain system. The largest of these pumps was the gasoline powered P-250 portable pump. The P-250 weighed over 150 pounds with fuel, and was capable of supplying 250 gallons per minute of firefighting water to either three 1 ½-inch hoses or a single 2 ½ inch hose. The P-250 could also be used to remove 250 gallons per minute of water from compartments. A smaller gasoline powered pump, the "handybilly," was also carried aboard naval vessels. The handybilly weighed 106 pounds and could supply firefighting water to a single 1 ½-inch hose or remove water at the rate of sixty gallons per minute. The handybilly could also be connected to a mechanical-foam nozzle to produce firefighting foam.

Naval vessels were equipped with numerous items designed to remove water from compartments. Portable electric submersible pumps could be dropped into a flooded compartment. A 2 ½-inch hose was connected to the pump discharge and carried water to the nearest available overboard discharge fitting. Overboard discharge fittings were fitted into the hull at frequent intervals to facilitate removal of firefighting and floodwater from internal compartments. They were usually located just above the ship's waterline, and were covered with watertight caps except while in use. These overboard fittings enhanced the efficiency of dewatering pumps by reducing the head pressure on the discharge side. If the discharge line from a portable pump were simply run overboard from the main deck, the higher head pressure would significantly reduce the pumping rate. For example, standard submersible pumps discharged 140 gallons per minute with a discharge head of seventy feet. If the discharge head was reduced to fifty feet, the same pump discharged 200 gallons per minute.[19]

Naval vessels carried an extensive array of eductors to remove water from internal compartments. These eductors varied widely in size and capacity, but all functioned on the same principal firefighting water was supplied to nozzles, or jets in the eductor body. As the water flowed through these jets, a vacuum was created in the eductor body. Water in the flooded compartment would be drawn up a suction line connected to the eductor body by this vacuum, and would mix with the firefighting water. This water mixture would then be discharged overboard. Fixed eductors were permanently installed in compartments and were fitted with permanent firemain supply, suction, and overboard discharge piping. Portable eductors could be carried where needed. They used firefighting hoses to supply water and carry water to overboard discharge connections.

Smaller eductors removed less than one hundred gallons of water per minute; larger eductors had a capacity of well over 1,000 gallons per minute.[20]

Other significant equipment carried aboard naval ships for controlling damage included tools to access locked or damaged compartments, such as bolt cutters, fire axes, and crowbars. Portable oxyacetylene cutting apparatus was used to cut holes in decks and bulkheads and to remove debris. Portable battery operated lanterns were invaluable, as were portable blowers and ducts to remove smoke and toxic gases from internal compartments.[21]

Personnel Protective Equipment

Protective gear was designed to reduce the hazards to crew members as they fought fires and damage aboard naval vessels. The protective gear available to *Forrestal's* crew was essentially identical to that used by US Navy sailors during the Second World War.

Uniforms worn aboard ship were designed to provide some protection against fire. Enlisted crew members wore cotton chambray shirts, dungaree pants, and steel-toed boots. Officers wore cotton khaki colored shirts and trousers and steel toed boots. During fires, crew members would button the top buttons on their shirts and tuck their trouser bottoms into their socks to minimize the amount of exposed flesh. However, the effectiveness of this procedure, which was already marginal, was reduced even more for the many crew members that frequently wore short sleeve shirts during warm weather. Personnel attacking the fire would also don asbestos gloves and helmets with a small attached battery operated lantern, known as a "miner's lamp."

Breathing apparatus was available to protect naval firefighters from hot, toxic gases. The most common type of breathing apparatus used for fighting fires on *Forrestal* was the "Oxygen Breathing Apparatus," or OBA. The OBA was a self-contained unit for individual firefighters. It consisted of a canister holder, two neoprene breathing bags (one on each side of the canister holder), a facepiece with inhalation and exhalation tubes, a timer, and a breastplate with webbing to attach the unit to the wearer. The firefighter wore the OBA on the front of his body. A fresh canister was inserted into the OBA before use. When activated, chemicals in the canister reacted with moisture from the firefighter's breath to produce oxygen and absorb carbon dioxide. The breathing bags held and cooled the oxygen. The firefighter manually set a timer to activate an audible alarm several minutes before the canister's chemicals were exhausted. The firefighter had to return to a clean atmosphere to change canisters. Each canister supplied approximately thirty minutes of oxygen.[22] Tending lines could be connected to the OBA to maintain lifeline signals with personnel remaining in safe atmospheres.

Aluminized asbestos "proximity suits" were carried aboard naval vessels. These protected personnel against high heat, but were not designed for direct contact with flames. Proximity suits were frequently used to rescue personnel, such as aircrew members involved in accidents on the flight deck.[23]

The preceding chapters have described the survivability features incorporated in warships operated by the US Navy in the 1960s, the damage control doctrine developed over the years, and the specialized damage control and personnel protective gear available to sailors. Although the "supercarriers" of the 1960s had dramatically increased in size and complexity compared with aircraft carriers that operated during World War

Two, the damage control tools available to sailors had not significantly changed. The Navy's World War Two damage reports clearly described the massive fuel, ordnance, and aircraft fires that occurred on carriers as a result of mishaps and enemy attacks, and *Forrestal* carried more aircraft, ordnance, and fuel than any aircraft carrier built before her. Unfortunately, her crew members were equipped with virtually the same equipment that their fathers had used to fight shipboard fires over twenty years earlier. This damage control equipment was not faulty or poorly designed; it had simply been rendered obsolete, and was not capable of quickly and effectively extinguishing a massive conflagration on the flight deck. The protective gear available to *Forrestal's* crew was woefully inadequate. Although the OBAs effectively protected firefighter's lungs, the non-fire retardant cotton uniforms worn by sailors provided virtually no protection against burns.

After the 1967 fire on *Forrestal*, the Navy took a hard look at the adequacy of damage control tools available to shipboard firefighters. Two investigations were convened shortly after this fire. The first of these focused solely on the *Forrestal* fire, but the second investigation examined the safety of aircraft carrier operations throughout the US Navy. These investigations developed numerous recommendations to improve shipboard damage control readiness. The next chapter examines the most significant of these proposed improvements.

---

[1] Jerome Williams, John J. Higginson, and John D. Rohrbough, *Sea and Air: The Marine Environment*, 2nd ed. (Annapolis, MD: UNITED STATES Naval Institute, 1975), 306-307.

[2] Thomas C. Gillmer, *Modern Ship Design,* 2nd ed. (Annapolis, MD: UNITED STATES Naval Institute, 1986), 88.

[3] G. C. Manning and T. L. Schumaker, *Principles of Warship Construction and Damage Control* (Annapolis, MD: UNITED STATES Naval Institute, 1935), 300-304.

[4] Bureau of Naval Personnel, *Principles of Naval Engineering* (Washington, D.C.: US Navy, 1970), 63-64.

[5] Ibid., 36.

[6] Ibid., 15-16.

[7] Kit Bonner and Carolyn Bonner, *Great Naval Disasters: U.S. Naval Accidents in the 20th Century* (Osceola, WI: MBI Publishing, 1998), 93.

[8] Norman Friedman, *U.S. Destroyers: An Illustrated Design History* (Annapolis, MD: UNITED STATES Naval Institute, 1982), 85, 118.

[9] Navy Department, *War Damage Report No. 56* (Washington, D.C.: US Navy, 1946), 22.

[10] Navy Department, *War Damage Report No. 51* (Washington, D.C.: US Navy, 1947), 18-19.

[11] *War Damage Report No. 56*, 26-29.

[12] Naval Ship Systems Command, *Naval Ships Technical Manual Chapter 9930: Fire Fighting – Ship* (Washington, D.C., Naval Ship Systems Command. 1967), 77-78.

[13] Ibid., 11-18.

[14] Ibid., 7-9.

[15] Ibid., 24-25.

[16] Ibid., 25-35.

[17] Ibid., 39-43.

[18] Ibid., 53.

[19] Ibid., 70.

[20] Ibid., 70-71.

[21] Ibid., 72-75.

[22] US Maritime Administration, *Marine Fire Prevention, Firefighting, and Fire Safety* (Washington, D.C., UNITED STATES Maritime Administration. 1987), 327-336.

[23] Ibid., 366-367.

CHAPTER 4

RECOMMENDATIONS FROM INVESTIGATING PANELS

Soon after the fire aboard USS *Forrestal*, two separate investigation panels were formed. The first of these investigations was required by naval regulations, and was conducted in accordance with instructions contained in the *Manual of the Judge Advocate General*. The purpose of the Judge Advocate General Investigation was to determine what caused the fire, and who was responsible. Rear Admiral Forsyth Massey headed this investigation, and produced a 7,500-page report containing the evidence he reviewed, along with his findings of fact, opinions, and recommendations. Admiral Massey found that serious deficiencies existed in *Forrestal's* damage control related design features. He also stated that the damage control and firefighting equipment carried aboard *Forrestal* was inadequate, and many members of *Forrestal's* damage control organization were poorly trained. His report included thirty-one proposals to correct these deficiencies.

The senior officer in the US Navy ordered the second of these investigation panels to be convened, shortly after Rear Admiral Massey's team began their work. Admiral Thomas Moorer, Chief of Naval Operations, appointed recently retired four-star Admiral James S. Russell as director of this panel. Admiral Russell was a former naval aviator, and had served as the Vice Chief of Naval Operations prior to his retirement. Admiral Russell was directed to examine aircraft carrier operations throughout the Navy, with the goal of assessing safety hazards and proposing ways to improve shipboard damage control effectiveness. Admiral Russell generally concurred with Admiral Massey's recommendations, and included them as proposed improvements in his report as well. However, Admiral Russell's report also included several proposals to improve personnel protective equipment available to shipboard personnel. Admiral Russell wrote that the Navy's available personnel protective gear was poor, and that more effective equipment was needed as soon as it could be developed.

This chapter will examine how these two panels conducted their investigations, the facts they discovered, the opinions they formed based on these facts, and the solutions they proposed to improve the deficiencies they perceived to exist.

The Judge Advocate General Investigation

Following the fire, the *Forrestal* steamed to Naval Air Station Cubi Point, Republic of the Philippines to conduct repairs. Although the scope of required repair work was too extensive to be accomplished at Cubi Point, inspections and basic repairs were made to ensure that *Forrestal* was able to safely return to the United States.

Vice Admiral Charles T. Booth, the US Atlantic Fleet Naval Air Force Commander, immediately ordered a *Manual of the Judge Advocate General* investigation into the *Forrestal* fire. Rear Admiral Forsyth Massey was appointed Senior Member of this Informal Board of Investigation on 30 July 1967. Rear Admiral Massey's primary assistants during the investigation were Captains A.K. Earnest and M.J. Stack. Commander Joseph H. Baum and Lieutenant Commander Edward T. Boywid provided legal counsel for the board. The members of the board arrived at NAS Cubi Point on 3 August 1967. The members began the investigation while temporary repairs were in progress, and remained aboard for the thirty-two-day transit back to *Forrestal's* homeport of Norfolk, VA.

Captain Beling's Immediate Superior in Command (ISIC), Rear Admiral Harvey P. Lanham, Commander of Carrier Division Two (COMCARDIVTWO), ordered his staff to conduct a preliminary investigation on 30 July. *Forrestal* was serving as Rear Admiral Lanham's flagship, and he and his staff were aboard during the fire. Rear Admiral Lanham's investigating team, headed by Captain William Morton, presented Rear Admiral Massey and his board with a background brief on the fire upon their arrival. Three members of COMCARDIVTWO's preliminary investigation team assisted Rear Admiral Massey's board throughout their investigation. These three officers included Commander Roger Carlquist, Commander Roger Weeks, and Ensign David Jacobs.[1]

The first significant task faced by Rear Admiral Massey's investigating board was the identification of "parties." The board members examined the duties and responsibilities inherent in billets of service members assigned to *Forrestal* during the fire. If the board determined that a service member's duties and responsibilities related to either the initiation of the fire or controlling the resulting damage, that serviceman was designated a party. Twenty personnel were designated as parties, and all were offered legal counsel. Rear Admiral Massey designated these parties shortly after his arrival to allow adequate time to embark desired legal counselors aboard *Forrestal* prior to the long transit back to Virginia.

After the parties were identified, the Investigating Board began taking statements from parties and witnesses. The board used formal hearing room procedures when taking statements, and all statements were taken under oath. During the investigation, the board read approximately 1,900 statements from 136 parties and witnesses.[2]

The investigating board also spent time touring the damaged areas of the ship and reviewed the Pilot Landing Aid Television (PLAT) camera film carefully. The PLAT camera was used to film all planes as they launched from or landed on the *Forrestal's* flight deck. When the fire began, the PLAT camera was filming a KA-3B aircraft as it prepared to launch. The camera recorded the accidental launch of the Zuni rocket. The PLAT operator then turned the camera and recorded the burning A-4 shortly after the Zuni rocket struck it. The camera's position was not changed again for the duration of the fire. The camera recorded the spread of the fire, the exploding ordnance, and the crew's firefighting efforts. The PLAT camera also recorded the time of these events by filming an integrated clock face. This footage proved invaluable to the investigators.[3]

Rear Admiral Massey submitted his investigation report to the commander of the US Atlantic Fleet Naval Air Force on 19 September 1967. The report consisted of approximately 7,500 pages, divided into thirteen volumes. Volume One contained the board's preliminary statement, findings of fact, opinions, and recommendations. The remaining volumes contained testimony and statements presented by witnesses.

Findings of Fact

The Investigating Board determined that the fire began at 10:52 a.m. local time on 29 July when a Zuni rocket struck A-4 aircraft number 405, puncturing its external 400-gallon fuel tank. A fragment also punctured the external fuel tank of nearby A-4 number 310. The burning fuel quickly spread to the after portion of the flight deck, pushed by thirty-two knots of wind and the exhaust of several jets positioned ahead of the stricken aircraft. General Quarters was sounded at 10:53 a.m., and material condition Zebra was set throughout the ship at 10:59 a.m. However, the crew left some Zebra fittings open to facilitate rapid evacuation of injured personnel.[4] The investigators found many of the high capacity foam and firefighting hoses on the port side of the flight deck were engulfed in flames and unusable. A 1,000-pound bomb fell from A-4 number 405 when it was struck by the rocket, and rolled into a pool of burning jet fuel. The casing of the bomb, which was split by the fall, quickly began to heat up. Fifty-four seconds after the fire began, Chief Petty Officer G.W. Farrier attempted to extinguish the burning pool of fuel around the bomb with a portable PKP dry chemical extinguisher. Approximately one minute and twenty seconds after the fire began, crew members attacked the forward boundary of the fire with firefighting water. One minute and thirty-four seconds into the fire, the first bomb exploded. This explosion killed Chief Farrier and twenty-six other fire fighters in the vicinity, and spread the fire to a group of three A-4 aircraft stationed near the after end of the flight deck. Several other hose teams continued to advance on the fire immediately after this explosion, but a second bomb exploded nine seconds after the first. The second bomb's explosion spread the fire to ten additional aircraft. Seven additional major explosions occurred in the next five minutes, severely hampering firefighting efforts on the flight deck.

Several of these explosions penetrated the armored steel flight deck and spread the fire to the three decks below the flight deck in the aft portion of the ship. The board determined that the burning aircraft contained a total of approximately 40,000 gallons of JP-5 fuel, and that this burning fuel spread the fire to the ship's sides, stern, and through holes in the flight deck into the hangar bay below. These bombs killed fifty night crew personnel who were sleeping in berthing compartments below the after portion of the flight deck. Forty-one additional crew members were killed in internal compartments in the after portion of *Forrestal*. The investigation found that firefighting foam and sprinklers effectively prevented the spread of fire in the hangar bay.

The investigators assessed the crew's firefighting efforts as effective after the nine major explosions subsided. "That once fire boundaries were established there was no further spread of the fire. Thereafter, the fire was fought aft progressively, compartment by compartment, on each deck in textbook fashion until it was finally extinguished. The only secondary damage was that caused by fire fighting water."[5] The flames on the flight deck were extinguished by 11:40 a.m., but fires in the internal compartments were not entirely extinguished until approximately 4:00 a.m. the morning of 30 July.[6] One hundred thirty-four crew members perished, and the fire and explosions injured 161 more. The estimated damage to the ship (not including damage to aircraft) was $72.1 million.[7]

Rear Admiral Massey's Board of Investigation dedicated a section of their findings to damage control and firefighting-related training, procedures, and material condition. First, the report stated that the normal damage control refresher-training period (REFTRA) was shortened from six weeks to four weeks for *Forrestal* prior to her deployment. Second, *Forrestal* received a grade of "unsatisfactory" in setting material condition Zebra during refresher training, but achieved a satisfactory grade during her predeployment Operational Readiness Inspection (ORI). Third, 37 percent of the ship's damage control personnel who attended refresher training transferred prior to *Forrestal's* deployment. At the time of the fire 1,610 crew members (57 percent of the ship's company) had attended firefighting school in the previous three years. Of course, this meant that 43 percent of the ship's company had not attended firefighting school in that time period. *Forrestal* conducted General Quarters drills fifty-seven times in the 106 days that she was at sea prior to the fire.[8]

The report also identified several fundamental training deficiencies that hindered firefighting efforts. The board found that numerous personnel on the flight deck were unfamiliar with firefighting procedures and equipment, and were unable to effectively contribute to firefighting efforts. For example, investigators discovered that at least one firefighting foam station was not initially charged because crew members were unsure how to activate the system. Rear Admiral Massey's team noted that the physical configuration and activation procedures varied considerably among *Forrestal's* different foam stations. This lack of standardization could easily prove confusing to sailors who were not thoroughly familiar with the foam generation stations. Another significant hindrance to effective firefighting efforts resulted because many crew members did not report to their assigned general quarters stations (some were unable to because of injuries, some were impeded by the ship's physical damage, some were already heavily involved in the firefighting efforts, and others simply made no attempt to reach their stations).[9]

The investigation report also noted several problems with Oxygen Breathing Apparatus (OBAs). "Significant numbers" of personnel assigned to Forrestal's air wing were not trained in using OBAs, some personnel experienced difficulty in activating the oxygen generating canisters in the OBAs, and some canisters did not last for the rated thirty-minute time period.[10]

Opinions and Recommendations

Rear Admiral Massey's report included 116 opinions based on the facts uncovered during the investigation. Many discussed the need to improve ordnance handling safety procedures, but a substantial number of opinions related to damage control. Although the report acknowledged several shortcomings in the crew's firefighting performance, it was particularly critical of the damage control equipment available aboard *Forrestal*:

> With existing installed fire fighting equipment, the fire could not have been extinguished prior to the explosion of major ordnance (ninety-four seconds after initiation of the fire) regardless of the aggressiveness, readiness, response and expertise of personnel and readiness of equipment…the design and operating procedures of fire fighting equipment currently available in attack carriers is totally inadequate to the needs generated by modern combat operations and the concentrations of very large quantities of ordnance and fuel on jet aircraft.[11]

The members of the board, based on their investigation into the fire, translated these opinions into sixty-two recommendations. Thirty-one of these recommendations were damage control related, and focused on improving training, damage control equipment, and warship design. To improve the performance of the shipboard damage control organization, the investigators recommended minimizing the transfer of trained personnel prior to a ship's deployment. This recommendation was especially pertinent since 37 percent of Forrestal's trained firefighters transferred from the ship prior to deployment.

Rear Admiral Massey also recommended that aircraft carrier air wing personnel receive increased firefighting and damage control training. Air wing personnel comprised nearly 40 percent of the deployed aircraft carrier's crew. These sailors operated and maintained the aircraft, and did not move aboard the ship until after the ship had completed a great deal of predeployment training. The air wings were not permanently attached to particular ships, and frequently deployed on different classes of aircraft carriers. As a result, the air wing sailors tended to be somewhat unfamiliar with the location and operation of firefighting and damage control equipment peculiar to the ship they were serving on. However, since these sailors primarily worked on and near the flight deck, it was essential for them to have a thorough understanding of firefighting techniques and equipment. The investigators specifically called for increasing instruction for air crew personnel in the following areas: shipboard damage control organization, principles of damage control, shipboard orientation (including traffic flow patterns during emergencies and escape routes, and how to activate and use damage control equipment such as OBAs, firefighting foam stations, the ship's firemain, and sprinkler systems.[12] The investigation report also recommended that all personnel assigned to aircraft carriers (including air wing personnel) achieve basic qualifications in damage control and firefighting prior to embarking their ships.

Rear Admiral Massey's team members also felt shipboard flight deck firefighting training drills were inadequate. They recommended that the Navy develop realistic training exercises based on fires of the magnitude experienced on *Forrestal*, simulating the hazards of live ordnance and the loss of key personnel and equipment.

As stated earlier, the investigating board believed that the fire on *Forrestal's* flight deck could not have been extinguished prior to the ordnance explosions with the equipment available onboard. To correct this unacceptable situation, the panel recommended that the Navy commission a study to examine improvements to increase the effectiveness of shipboard firefighting and damage control equipment. Specifically, the report recommended that this study focus on potential improvements to firefighting foam stations, firefighting nozzles, and fire hose storage.

Recommended improvements to foam stations included standardizing controls to reduce operator confusion. The investigation had discovered that the operating controls varied with the different foam stations located throughout the ship. This lack of standardization was especially confusing for members of the embarked air wing, who were often unfamiliar with a particular ship's equipment idiosyncrasies. The report also recommended increasing the number of remote activation controls for each firefighting foam station to improve response time. Testing completed during the course of the investigation revealed significant delays between activation of the foam stations and delivery of firefighting foam to the flight deck hoses. Investigators tested the performance of ten foam stations without providing advance warning to the *Forrestal's* crew. One station produced foam after seventeen seconds had elapsed, another station failed to develop foam at all, and one station generated foam after four minutes. The remaining seven stations produced firefighting foam thirty to forty-five seconds after they were activated.[13] Since the first bomb exploded on *Forrestal's* flight deck one minute and thirty-four seconds into the conflagration, investigators recommended that the Navy examine the feasibility of modifying the foam stations to reduce the time required to deliver foam to flight deck hoses.

As the *Forrestal's* crew battled fires that had spread into compartments below the flight deck, they were forced to cut small access holes into several bulkheads and decks to insert nozzles and hoses. This technique proved useful in cooling compartments to prevent the spread of fire, and fighting fires where the normal entry points were inaccessible because of damage or high intensity fires. The panel recommended that the Navy develop and issue small omni-directional nozzles, especially designed to spray all areas within a compartment when inserted through a small hole in a bulkhead or deck.

Rear Admiral Massey's investigators discovered that the heavy firefighting hoses used on the flight deck were very susceptible to getting tangled up as they were deployed. If a hose developed a significant kink while being used to fight fire, the flow of water or foam would be interrupted. The sudden loss of agent would render the hose ineffective until the kink was removed, and could easily endanger firefighters if they were in close proximity to a large fire. The report recommended that the Navy study ways of improving hose storage to reduce tangling during hose deployment.

Rear Admiral Massey also proposed significantly increasing the allowance of firefighting foam, OBAs, and OBA canisters carried aboard *Forrestal*. The board opined that the existing allowance of foam and OBA canisters was insufficient for combating serious fires, and believed that *Forrestal's* crew would have been forced to simply contain the fires until they burned out if other ships in the vicinity had not replenished these items. *Forrestal's* existing allowance included 1,220 five-gallon containers of firefighting foam concentrate, 550 OBAs, and 3,300 OBA canisters. The board recommended increasing this allowance to 2,500 containers of foam, 620 OBAs, and 8,000 OBA canisters.[14]

Rear Admiral Massey also recommended that the Navy consider employing armored fire fighting vehicles on the flight decks of aircraft carriers. The report noted that such vehicles would provide carriers with several useful capabilities. They could be used to push burning wreckage (such as damaged planes) over the side, they could closely approach fires while protecting operators from the hazards of ordnance detonation and resulting shrapnel, and supervisors could direct their employment by radios.

Finally, the initial investigation report into the fire on *Forrestal* recommended several modifications to the Navy's carriers to improve survivability and enhance the damage control efforts of crew members. The report noted that approximately 40,000 gallons of fuel from burning aircraft contributed significantly to the intensity of the fire.

The burning fuel also entered interior compartments through bomb holes and other opening in the flight deck, spreading the fire and damage. Rear Admiral Massey recommended that the Navy add large sprinkler systems specifically designed to quickly wash large quantities of fuel off carrier flight decks. He noted that a large system of drains would have to be added as well to accommodate large volumes of fuel and water. These drains would have to be designed to divert fuel and water over the side while minimizing fuel intrusion into interior compartments. The board also recommended extending the length of flight decks over the stern of aircraft carriers to eliminate another potential route for burning fuel to enter the ship. Finally, the board recommended incorporating jettison ramps into the flight deck so that ordnance, flammable materials, and even aircraft could be quickly pushed over the side when necessary.

During the *Forrestal's* fire, ninety-one crew members died in compartments below the flight deck. Some crew members were trapped in compartments because the explosions damaged a single exit. Others died because they were unable to reach the nearest exit before toxic gases and heat overcame them. To reduce similar casualties in the future, the board recommended that the Navy construct alternate escape exits in compartments of all vessels, where possible.

Numerous crew members stated that the shipwide general announcing system, the "1MC," was nearly impossible to hear in the hangar bay during the fire. This announcing circuit was critical, since senior officers frequently used it to provide direction and status updates to the crew during emergencies. Testing by the investigators confirmed that the system was unintelligible throughout much of the hangar bay, so they recommended that this deficiency be corrected.

While Rear Admiral Massey's team was crossing the Atlantic and continuing their investigation, the Chief of Naval Operations, Admiral Thomas Moorer, decided to establish a panel to review the safety on aircraft carriers throughout the Navy. As discussed earlier, Admiral Moorer selected recently retired former Vice Chief of Naval Operations and naval aviator Admiral James S. Russell to head this panel.

The Russell Report

Admiral Russell's panel convened in Washington, D.C. on 15 August 1967, just over two weeks after the *Forrestal's* fire. In addition to Admiral Russell, who served as the Director, the Office of the Chief of Naval Personnel assigned eleven officers and civilians to this Panel to Review Safety in Aircraft Carrier Operations (PRSACO). These panel members were selected based on their professional expertise and experience with aircraft carrier operations and equipment design. The PRSACO members conducted a series of organizational meetings during their first five days together, then split into two groups. The first group was comprised of Admiral Russell, Rear Admiral Buie, Captain McCall, Commander Engel, Commander Charles, and Mr. Bee. This group visited the headquarters of the Pacific Fleet's Commander in Chief and spent a week assessing four aircraft carriers as they conducted combat operations in the Gulf of Tonkin, off Vietnam.

While Admiral Russell's group was conducting its tour, the remaining panel members conducted a review of available literature on the topic. When Admiral Russell returned from his tour, the entire panel reconvened in San Diego, California. The panel then conducted conferences with personnel serving on the staff of the Commander, Naval Air Forces Pacific (COMNAVAIRPAC) and the Pacific Training Command (COMTRAPAC). After these conferences, the panel members returned to the Pentagon for a series of briefings and discussions. Top Navy leaders considered the briefings presented to Admiral Russell's panel important. The Naval Material Command, the Bureau of Naval Personnel, and the Office of the Chief of Naval Operations produced the majority of briefings. The Chief of Naval Operations wrote letters to the Chiefs of Naval Material and Personnel requesting briefings on subjects of interest to the panel. He also wrote internal memorandums directing his staff to provide desired briefings for Admiral Russell's panel. The members of the panel received seventy-six separate briefings. A short description of each of these briefings was included in the panel's report. During their review of carrier operations safety, PRSACO members studied a bibliography of eighty-one pertinent books, articles, and reports. Items in this bibliography discussed ordnance safety, personnel issues such as training, organization, and personal protective equipment; damage control doctrine, World War Two battle damage reports, reports of fires on aircraft carriers after the Second World War, and other ship systems. A brief synopsis of each item in the bibliography was included in Admiral Russell's report. The *Russell Report* (as Admiral Russell's *Report of the Panel to Review Safety in Carrier Operations* was frequently referred to in Navy memorandums) included a list of forty-six interviews that panel members conducted. This extensive series of interviews included discussion with the Navy's top leaders and carrier aviation experts, including the Secretary of the Navy and thirty-nine separate flag officers (admirals). Panel members interviewed several of these senior leaders on more than one occasion. Most of them had extensive experience with aircraft carrier operations, and several were former commanding officers of carriers. The positions of these senior leaders were diverse -- some were responsible for training, many supervised technical research and ship construction programs, and others commanded operational units. The panel members interviewed Rear Admiral Massey to discuss insights he gained while investigating *Forrestal's* fire. They also interviewed the former Commanding Officer of *Forrestal*, Captain John Beling, and *Forrestal's* Chief Engineer, Commander Merv Roland. Finally, the panel visited *Forrestal* while it was docked in Norfolk, Virginia to examine the damage caused by the fire.

Admiral Russell's team completed their *Report of the Panel to Review Safety in Carrier Operations* on 16 October 1967. In the abstract, Admiral Russell stated that his review identified several serious deficiencies:

> Deficiencies were identified, however, that, though largely beyond the ability of the ships to correct, do affect the ability of CVAs (aircraft carriers) to carry out their assigned combat missions with a reasonable degree of safety to themselves. The most serious of these deficiencies are inadequate fire protection for the flight deck and to a lesser extent the hangar deck. . .personal equipment for fighting fires and for individual survival,…inadequate individual and team training.[15]

Admiral Russell provided eighty-six recommendations to improve aircraft carrier safety in his report. Many of these recommendations, as in Rear Admiral Massey's report, were focused on improving damage control training, increasing the capability of shipboard firefighting and other damage control equipment, and modifying warship design to further enhance survivability. However, it is important to note that Admiral Russell's panel had a much broader point of view than Rear Admiral Massey's team. Rear Admiral Massey was appointed to determine what caused the fire on *Forrestal*. His recommendations were based on facts discovered during his investigation and opinions that he formed after closely studying that single incident. Rear Admiral Massey had at least some incentive to moderate his recommendations – any harsh criticism of the performance of *Forrestal's* crew members would be damaging to the careers of officers serving on *Forrestal* (Rear Admiral Massey specifically stated in his report that he did not recommend placing blame on any *Forrestal* crew members for the conflagration). On the other hand, the highest ranking officer in the United States Navy, Admiral Moorer, appointed Admiral Russell to undertake a comprehensive study of how safely all American aircraft carriers were being operated. Although his project was important to the Navy, Admiral Russell was not as constrained by the need to quickly complete an investigation before eyewitness accounts of a single disaster deteriorated. Admiral Russell's panel visited five aircraft carriers to gain firsthand knowledge of how they were operated. The Navy's leading experts on firefighting and damage control, warship design and construction, and procurement briefed the PRSACO members. They were granted virtually unlimited access to interview Navy uniformed personnel and Department of the Navy civilians to capitalize on their tremendous experience and gain insight from their opinions. A final significant factor was that Admiral Russell's panel was well designed to assure a thorough, honest look at existing flaws in aircraft carrier operations safety. Admiral Moorer's decision to appoint a retired officer avoided the possible negative career implications an active-duty officer might face if he chose to advocate politically unpopular recommendations. Each panel member was allowed and encouraged to present possible recommendations to be considered for inclusion in the final report. However, only Admiral Russell had authority to approve what recommendations were included in his report. This enabled panel members to present honest and critical recommendations without concern for their own careers. Additionally, the presence of a retired four-star admiral on the panel (with obvious strong support from the Chief of Naval Operations) helped ensure that the panel would receive a high degree of cooperation from the busy officials they chose to interview.

Each of the eighty-six recommendations proposed by Admiral Russell was accompanied by an explanation of why they were considered important by the panel. The panel used information gathered from the sources mentioned earlier to justify their recommendations. Admiral Russell included proposals to assign cognizance for each of his recommendations to a specific naval command. For example, he proposed that the Naval Material Command be assigned responsibility for improving fire hoses used aboard ships. Three days after Admiral Russell submitted his report, the Chief of Naval Operations forwarded the report to an extensive distribution list of naval commands.

Admiral Russell's panel grouped their recommendations into nine separate categories--ship's material, personal equipment, aircraft systems, weapons, training, documentation, personnel, organization, and operations. This thesis will not examine those recommendations related to aircraft systems, weapons, or documentation (since the recommendations in this category related to ordnance safety publications). The recommendations in the remaining six categories that pertain to damage control improvement will be examined.

Recommendations to Improve Warship Survivability Features

In his most significant recommendation for improving ship's material, Admiral Russell proposed developing advanced flight deck fire fighting systems for carriers. Many of the elements of this proposed system were initially included in Rear Admiral Massey's report, such as remote control, rapid response time, and massive firefighting agent delivery capability. This *Russell Report* recommendation also incorporated another consequential proposal from Rear Admiral Massey's initial report--that it was important to incorporate means for quickly draining large quantities of spilled fuel from flight decks into this advanced fire fighting system.[16] Admiral Russell provided strong supporting rationale for including this recommendation in his report:

> Development of an advanced flight deck fire fighting system is of prime importance. Principal attention in aircraft carrier fire fighting has been focused in the past on the hangar deck. Adoption of the steel ballistic deck in *Midway* Class and later carriers, and the trend toward elimination of aviation gasoline, tended to support the belief that control of fire on the flight deck was not a serious problem. The *Forrestal* incident proved that it is. Modern carrier aircraft are capable of carrying large quantities of fuel and weapons. The strike group on *Forrestal* was estimated to be loaded with approximately 40,000 gallons of JP-5 (jet fuel) when the accident occurred. Modern aircraft and weapons complexities combine with environmental conditions on a flight deck to provide an ever-present possible source of ignition. Presently installed equipment is not capable of handling a conflagration of the magnitude of that which developed on *Forrestal*.[17]

Admiral Russell provided three pages of justification detailing why he considered it vitally important to develop an advanced firefighting system for carrier flight decks. He concurred with Rear Admiral Massey's assessment that existing flight deck fire fighting equipment was simply inadequate, and proposed that the Naval Material Command immediately begin research and development on an improved system.

The *Russell Report* recommended that the Naval Material Command develop a standardized system for marking and illuminating emergency escape routes from interior compartments. The report noted that personnel experienced difficulty in escaping from smoke-filled compartments in many previous shipboard fires, including the one on *Forrestal*. PRSACO members believed two critical factors increased the difficulty crew members experienced when attempting to evacuate dark, smoke filled compartments during emergencies: aircraft carriers were extremely large, and many crew members were not familiar with all sections of their ships. During their visit to four deployed carriers, panel members observed various markings designed to help personnel evacuate compartments during emergencies; they also noted that some carriers had no markings at all. The panel recommended that the Naval Material Command investigate which colors and types of paint and lights were most effective in helping personnel evacuate shipboard compartments, and then develop an effective, standardized system to mark evacuation routes.[18] The PRSACO members proposed implementing a closely related recommendation originally made by Rear Admiral Massey, increasing the number of exits from compartments. They noted that US Navy ship design specifications required two exits from all stations normally manned by ten or more crew members. However, this requirement did not apply to berthing compartments or workshops. Admiral Russell's team proposed extending this requirement to include all berthing compartments and working areas that were sometimes occupied by ten or more crew members. They recommended that all new ships be constructed to this standard, and that existing ships be altered to meet this new requirement where feasible.[19]

Admiral Russell also concurred with the assessment of *Forrestal* fire investigators that the general announcing system was difficult to hear and understand in some locations on aircraft carriers. The investigation into *Forrestal's* fire determined that personnel in the hangar deck directly below the flight deck had difficulty hearing and understanding the ship's general announcing system, or "1 MC." Admiral Russell's report noted that this problem was not confined to *Forrestal*:

> The complaint concerning the inadequacy of the 1 MC General Announcing System is universally supported by Yankee Station CVAs visited by the Panel. There are many working and living areas where the 1 MC simply cannot be heard. . . . A space-by-space survey should be conducted on each ship in order to determine what must be done to provide a system that will be adequate

for passing important information or orders during an emergency. It is equally important to insure that the General and Chemical Alarms may be heard in every normally inhabited space throughout the ship and that the inhabitants of these spaces may also communicate the existence of a fire or other emergency in the space to the bridge. In short, command cannot function properly without adequate command and control communications.[20]

Remembering that fifty crew members died while sleeping in their berthing compartments after the General Alarm had been sounded over *Forrestal's* 1 MC system, it is hard to overemphasize the importance of this recommendation. Admiral Russell proposed that the Naval Material Command should conduct surveys on all US Navy aircraft carriers to discover and correct instances where 1 MC speakers were inaudible or unintelligible in inhabited compartments.

Admiral Russell recommended that the Naval Material Command review all pending repair requests for the Navy's carriers. He proposed that all items affecting fire fighting or damage control should be considered critical to the safety of these ships, and should be given high priority during each ship's maintenance periods. His included rationale noted that there is always limitations on how many items can be repaired or upgraded during any given maintenance period, and there is heavy competition about which items are given priority. Damage control improvements and repairs competed with areas such as propulsion machinery and command and control equipment. Admiral Russell stated that damage control items often lost out in this competition for limited resources:

> This fact, combined with Navy-wide apathy toward damage control over the past years, has resulted in the low state of material readiness in this important area.[21]

Admiral Russell cited excerpts from Inspector General assessments of five aircraft carriers conducted in May and June 1967 to support his harsh comment:

Fire fighting equipment in 3 of the 5 carriers inspected was in satisfactory or better condition. An examination of watertight inspection records and watertight boundaries revealed: watertight inspections were logged in 4 of 5 carriers; watertight boundaries were unsatisfactory in all 5 ships.[22]

PRSACO members believed that placing damage control repair requests in the "safety to ship" category would highlight their importance to senior officials, and help ensure that they received a higher priority in future maintenance periods.

The final significant recommendation by Admiral Russell's panel related to ship's material proposed establishing an exploratory program to study means of improving survivability of ships. To bolster this recommendation, the report stated that the Navy had great need for such a program: "As an example of the need, present-day shipboard fire fighting and damage control are essentially based on means available in World War Two."[23]

The panel theorized that this program could incorporate computer simulations to model damage that could occur from both accidents and enemy action, and that computers could also be used to evaluate the effectiveness of proposals designed to increase warship survivability.

<u>Recommendations to Improve Personal Protective
and Damage Control Equipment</u>

The need for improved personal protective equipment was the one area where Admiral Russell's report made significant recommendations not originally proposed by Rear Admiral Massey's team. The PRSACO members, based on their broader view, realized that shipboard emergency personal protective equipment was woefully inadequate. As Admiral Russell stated in the conclusion to his report:

> Of great importance in the handling of emergencies resulting from fire and explosion on a carrier is the personal equipment available for use in combating

the situation and in individual survival in a smoke/fire environment. Not much improvement has been made in these equipments since World War II. Major improvements are required and, with the advanced technology now available, these improvements should not be too difficult.[24]

Perhaps the most important personal protective equipment recommendation submitted by PRSACO members was the need for a more effective escape-breathing device. At the time of the *Forrestal's* fire, crew members often wore gas masks as they attempted to escape smoke-filled compartments during shipboard fires. These gas masks, primarily intended to protect crew members against attacks from chemical weapons, also provided some protection against hot smoky environments--they filtered solid particulate matter from the air, reduced the temperature of the air slightly, and served as a heat shield for the wearer's face. However, they provided no protection against toxic gases such as carbon monoxide, and were worthless to the wearer if a compartment's oxygen level was depleted by fire. The Navy's *World War II Damage Reports* found that the gas masks were often useful to personnel evacuating smoke-filled interior compartments: "On the basis of the service experience it is apparent that Navy Service Gas Masks are reasonably effective against smoke. Personnel must be thoroughly acquainted with their limitations, however."[25]

Admiral Russell concurred with this assessment, and recommended that the Naval Material Command distribute information throughout the Navy explaining the capabilities and limitations of the gas mask when used as an escape breathing device. He also stated that gas masks had been issued to the crew of only one of the carriers he observed off the coast of Vietnam. The gas masks of the remaining three carriers were stowed in storerooms, inaccessible to their crews in the event of sudden emergencies.[26]

Although he agreed with the twenty year old *World War II Damage Reports* that gas masks provided useful protection to personnel evacuating smoke-filled compartments, Admiral Russell felt that a more effective device was needed. He noted that personal emergency air masks were available to US Navy submarine sailors, and proposed that the Naval Material Command modify gas masks by adding small portable air cylinders. These cylinders would supply gas mask wearers with clean air for several minutes and increase their chances of escaping from compartments filled with toxic gases.

Admiral Russell also proposed that the Naval Material Command establish a program to improve the OBAs used by shipboard firefighters. His report stated that the OBAs were excellent tools, but noted that several deficiencies had been discovered during fires on naval vessels. Rear Admiral Massey's investigators discovered that many of the OBA canisters used to combat *Forrestal's* fire did not last the rated thirty minutes. *Forrestal* fire investigators also believed that many parts of the OBA were susceptible to deterioration over time, and were subject to breakage as a result of rough handling. The investigation team that studied a major fire on the carrier USS *Oriskany* had reached similar conclusions in 1966. PRSACO members suggested that research could result in OBA canisters with longer lives, and that design improvements could produce smaller, simpler, and more rugged OBAs.[27]

Finally, Admiral Russell proposed improving the clothing worn by personnel responding to fires on flight decks. He recommended upgrading the proximity suit used by sailors to rescue personnel from burning aircraft by improving its resistance to wear and tear, making it more flexible, and increasing its ability to reflect heat. He noted that several research reports confirmed that vastly improved aluminized fabrics were available and could be used to improve proximity suits.[28] He also noted that the jerseys worn by personnel working on the flight deck were not flame retardant, and their shoes were soft-toed and had poor treads on their soles. The *Russell Report* cited research demonstrating that it was feasible to treat clothing to make it flame retardant, and proposed that the Navy issue flame retardant clothing and improved footwear to shipboard sailors.

The *Forrestal* fire investigation team originally proposed many of the improvements to shipboard damage control equipment Admiral Russell recommended. For example, building on a recommendation made by Rear Admiral Massey, the *Russell Report* proposed improving shipboard fire hoses. The *Forrestal* investigators discovered that the cotton-jacketed hoses used throughout the ship were very susceptible to becoming tangled. PRSACO members confirmed this during their visits to four operational carriers, and also noted that the cotton-jacketed hoses were quickly worn out by being dragged across abrasive decks during training drills. The *Russell Report* described a project where improved hoses were being tested aboard another carrier, USS *America*. *America* had tested 235 lengths of neoprene-wrapped hose, and found that these new hoses did not tangle up and were significantly more wear-resistant than the cotton-jacketed hoses. Admiral Russell recommended that all aircraft carriers be equipped with neoprene-wrapped hoses in their hangar and flight decks. He also recommended that the Naval Material Command develop quick-disconnect couplings for these hoses to facilitate rapidly adding sections of hoses when needed.[29]

In a similar vein to his recommendation that repairs to damage control equipment be given a high priority in the competition for limited resources, Admiral Russell proposed that a portion of each ship's operating funds be allocated solely for the purchase of damage control and firefighting equipment. His supporting rationale clearly illustrated the many items that competed for funding priority and provided strong justification for why he considered this recommendation important:

> Operating funds are allocated to individual ships in the form of and Operating Target (OPTAR) by the type commander. Normal practice is for the ships to further sub-allocate amounts to each department. Out of each OPTAR must be obligated funds for such things as spare parts, consumables such as paper and soap, maintenance items such as wire and sheet metal, habitability items such as paint and deck tile, and replacement of equipage such as worn out fire hose and lost battle lanterns. The amount of the OPTAR is never enough to cover all of a ship's operating needs. Normal practice is to establish a priority list and fund down the list to the point where money runs out. In this system, the completeness of the inventory and the good material condition of damage control equipment must compete with all other consumables, spares, and equipage replacement, for funds. The tendency has been in the recent past, to place damage control gear low on the priority list. This has meant that inventories and material condition of damage control equipment were generally poor.[30]

In addition to establishing a separate pool of money for damage control equipment, the *Report of the Panel to Review Safety in Carrier Operations* recommended the Naval Material Command conduct further analysis of shipboard fires to determine a more appropriate allowance of OBAs and their canisters, fire fighting foam, fire extinguishers, and hoses. The report noted that both *World War II Damage Reports* and more recent investigations following major shipboard fires recommended significantly increasing the number of OBAs and OBA canisters.[31]

Recommendations to Increase Damage Control Awareness and Training

The remaining recommendations made by Admiral Russell's panel emphasized the urgent need to improve the level of damage control awareness and training throughout the fleet. The foremost recommendation Admiral Russell proposed to alleviate this situation was for the Chief of Naval Operations to ensure that air wing personnel received damage control training prior to deploying with an aircraft carrier. His report noted that sailors in the air wing comprised approximately 40 percent of the personnel on deployed carriers, and that an even higher percentage of air wing sailors were involved in fire fighting efforts on the *Forrestal* because most of them worked in the vicinity of the flight deck. However, his report was highly critical of the effectiveness of these sailors during the conflagration:

> Many of these air wing personnel, despite their courageous acts and strong desire to help, were ineffective and in some cases a hindrance to the fire fighting effort. These men had received no formal training in fire fighting or the principles of damage control. During a carrier's refresher training period, which is primarily devoted to ships damage control training, the air wing is not aboard, and no substitute damage control training is provided.[32]

Admiral Russell observed that requirements did exist for these sailors to receive damage control training. However, he found that the requirements were not being met for a variety of reasons, such as insufficient school capacity, high personnel turnover rates, lack of realistic training aids, insufficient attention by commanding officers, and insufficient requirements for ships to conduct periodic drills. His report included several proposals designed to alleviate these shortfalls.

Based on briefings he received from the Commanders of the Atlantic and Pacific Naval Air Forces and the Commanders of the Atlantic and Pacific Training Commands, Admiral Russell recommended increasing the throughput capability of damage control training schools by adding more instructor billets. He also proposed sending damage control training teams to assist deployed ships. The briefings presented to Admiral Russell indicated that vastly increased student throughput was required to meet existing training requirements. Officials estimated that school capacity was only sufficient to meet approximately 60 percent of the training requirements for the Pacific Fleet, and approximately 32 percent of the Atlantic Fleet's training requirements.[33] However, even this meager capability was severely underutilized, as the *Russell Report* clearly shows:

> COMNAVAIRPAC requires that the executive officer, and all repair party personnel attend a five-day fire-fighting course and all other personnel, including the Air Wing attend the two-day course…During FY '67 only 226 Air Wing personnel were trained. COMNAVAIRLANT requires that all repair party and in-port firefighting party members attend the five-day fire-fighting course, all air department personnel attend a three-day course, and half the ship's company attend a two-day fire-fighting course. In FY '67, no air-group personnel attended basic or refresher fire-fighting courses.[34]

The impact of these depressing macrolevel statistics were evident in the investigation reports of shipboard fires, which concurred that a dire need for increased damage control training existed:

> About 25 percent of the *USS Oriskany* crew and apparently none of the Air Wing personnel had received fire-fighting training prior to the October 1966 fire. Only 150 personnel were trained in the use of the OBA. On *USS Forrestal* about 50 percent of the crew and none of the Air Wing personnel had fire-fighting training prior to the fire. Both reports of these incidents recommended full-crew training in fire-fighting.[35]

To increase the awareness of the importance of damage control training on aircraft carriers, Admiral Russell's report recommended incorporating damage control training into the precommand training pipeline given to aircraft carrier commanding officers. His report noted that commanding officers of carriers were aviators with little or no prior damage control training or experience, and speculated that this could result in decreased command emphasis on the importance of damage control:

> This lack of experience in damage control on the part of the commanding officer is most critically reflected in a generally low level of command interest in damage control matters, and a failure to appreciate the importance of damage control training. Regardless of the enthusiasm and ability of the DCA (Damage Control Assistant), ship-controlling drills in damage control are not going to be included in an already-too-full schedule, unless the commanding officer recognizes the importance of damage control and the necessity for continued damage control training.[36]

Admiral Russell recommended that the training provided to future carrier commanding officers should include instruction on the principles of damage control, review of significant previous shipboard fires and battle damage, and participation in fire fighting and damage control training exercises. His report also proposed that newly enlisted personnel receive damage control training prior to reporting aboard, and recommended that officer-commissioning programs increase their emphasis on damage control training. His report stated that a decision had been recently made to eliminate the sole course on damage control principles included in the Naval Academy's curriculum, and he strongly advised reversing that decision.[37]

Finally, the *Russell Report* recommended that the Naval Material Command create improved damage control training aids for shipboard personnel. He proposed incorporating the PLAT camera footage into a training film to give shipboard firefighters a sense of the magnitude of fires they could encounter. He also proposed developing reusable training canisters for OBAs, so that ships could conduct OBA familiarization training without decreasing the amount of canisters available during actual fires.

As the preceding chapter illustrated, Admiral Russell's *Report of the Panel to Review Safety in Carrier Operations* proposed a plethora of possible means to improve damage control and firefighting capability on US Navy ships. Many of his suggestions incorporated recommendations originally included in Rear Admiral Forsyth Massey's investigation into the *Forrestal's* fire. The next chapter examines how the Navy implemented these important recommendations.

---

[1]Department of the Navy, *Manual of the Judge Advocate General Basic Final Investigative Report Concerning the Fire on Board the USS Forrestal (CVA-59) on July*

29, 1967. (Washington, D.C.: US Navy Office of the Judge Advocate General, 1968), 1-2.

[2]Ibid., 6.

[3]Ibid., 34.

[4]Ibid., 35.

[5]Ibid., 38.

[6]Ibid.

[7]Ibid., 33.

[8]Ibid., 77.

[9]Ibid., 81.

[10]Ibid., 83.

[11]Ibid., 112.

[12]Ibid., 129.

[13]Ibid., 82.

[14]Ibid., 126.

[15]Admiral James S. Russell, *Report of the Panel to Review Safety in Carrier Operations* (Washington, D.C.: Office of the Chief of Naval Operations, 1967), I-1.

[16]Ibid., A-1.

[17]Ibid., A-1.

[18]Ibid., A-12.

[19]Ibid., A-36.

[20]Ibid., A-15.

[21]Ibid., A-17.

[22]Ibid., A-18.

[23]Ibid., A-47.

[24] Ibid., V-1-V-2.

[25] Ibid., A-50.

[26] Ibid.

[27] Ibid., A-57.

[28] Ibid., A-58.

[29] Ibid., A-16.

[30] Ibid., A-33.

[31] Ibid., A-32.

[32] Ibid., A-73.

[33] Ibid., A-74--A-75.

[34] Ibid., A-74.

[35] Ibid., A-75.

[36] Ibid., A-76.

[37] Ibid., A-80.

CHAPTER 5

IMPLEMENTATION OF REPORT RECOMMENDATIONS

As the previous chapter discussed, two significant investigation panels convened shortly after the July 1967 fire aboard USS *Forrestal*. Both of the reports produced by these panels included numerous proposals to improve the effectiveness of damage control efforts on US Navy ships. However, Admiral Russell's broader investigation into the safety of carrier operations throughout the Navy appears to have had greater impact on damage control improvements in the US Navy. This chapter examines how the Navy implemented these recommendations included in the *Basic Final Investigative Report Concerning the Fire on Board the USS Forrestal* and the *Report of the Panel to Review Safety in Carrier Operations*.

Implementation of *Forrestal Fire Investigative Report's* Recommendations

The first of these panels, headed by Rear Admiral Forsyth Massey, conducted an investigation into the fire following the Navy's *Manual of the Judge Advocate General*. This type of investigation was commonly referred to as a "JAGMAN" investigation within the Navy. The primary purpose of a JAGMAN investigation was to determine the causes of an accident, and who should be held responsible for the resulting damage. Rear Admiral Massey did this, but he also provided thirty-one recommendations aimed at improving damage control deficiencies he observed during his investigation. The preceding chapter discussed fourteen of the most significant recommendations proposed by this investigation. Rear Admiral Massey completed his investigative report on 19 September 1967, and submitted it to Vice Admiral Charles T. Booth, Commander of the US Atlantic Fleet Naval Air Force, for review. Vice Admiral Booth approved the vast majority of recommendations proposed by Rear Admiral Massey, dissenting with only two of the damage control related proposals. The first of these two not approved recommendations had proposed minimizing the transfer of trained personnel prior to a ship's deployment. To justify their proposal, the investigation team noted that 37 percent of *Forrestal's* trained firefighters transferred prior to her deployment, and opined that these transfers had a significant negative impact on *Forrestal's* overall damage control readiness.[1] Vice Admiral Booth's endorsement letter on the investigation stated that high personnel turnover rates were common throughout the fleet because relatively few sailors assigned to aircraft carriers were re-enlisting after the expiration of their terms of required service. He further emphasized his point by stating that:

> These (fleet manpower) resources are not adequate to the task of stabilizing ship and squadron personnel from commencement of refresher training to completion of deployment. Indeed, when two or three aircraft carriers are scheduled to deploy in a two or three month time frame, fleet manpower resources are hard put to provide even the minimum manpower requirements.[2]

In short, although he had no objection to the concept of stabilizing manning on aircraft carriers, Vice Admiral Booth did not believe the Navy had sufficient manpower available to make this idea feasible.

Vice Admiral Booth also decided against immediately increasing the allowance of OBAs, OBA canisters, and firefighting foam concentrate carried aboard aircraft carriers. Rear Admiral Massey's team had proposed increasing the allowance of foam concentrate from 1,220 five-gallon cans to 2,500; increasing the number of OBAs from 550 to 620; and increasing the number of OBA canisters from 3,300 to 8,000. His report noted that *Forrestal* received substantial quantities of these items from other US Navy ships in her vicinity during her fire, and stated that he believed it would have taken significantly more time to extinguish the blaze without those supplements.[3] Admiral Booth did not completely discount this proposal, but he decided that detailed analysis was required prior to increasing allowance of these items. His letter stated that this analysis would have to include the increased cost of constructing stowage facilities for these items. Admiral Booth recommended delaying implementation of this recommendation even if higher authority decided to increase the allowance of these items until additional dedicated funds could be budgeted for these items.[4]

In his two-page long endorsing letter, Vice Admiral Booth praised the thoroughness of the report and the worth of recommendations presented by the investigating board. He noted that since the report contained so much important information, he was forwarding complete copies to the Commander in Chief of the Pacific Fleet, the Commander of Naval Air Forces in the Pacific, and the Seventh Fleet commander (under whose control carriers operated while prosecuting the war in Vietnam). He also forwarded excerpts of the report containing the investigation board's findings of fact, opinions, and recommendations to all Carrier Division commanders in the Atlantic Fleet. Vice Admiral Booth completed his review of the report on 26 September 1967 and forwarded it to his boss, Admiral Ephraim P. Holmes, Commander in Chief of the US Atlantic Fleet.

In contrast to Vice Admiral Booth's quick review of the report, which only lasted one week, Admiral Holmes took approximately two months to analyze the contents of Rear Admiral Massey's report. Admiral Holmes did not complete his endorsing letter until 1 December 1967. Admiral Holmes's eight-page endorsing letter was much more critical of the investigative report than that of Vice Admiral Booth. Admiral Holmes disagreed with the investigation board's assessment that the fire and resulting deaths and destruction were not the fault of any of Forrestal's crew members:

> The Commander in Chief U.S. Atlantic Fleet, therefore, specifically does not concur in Opinion 115 of the *Report of Investigation* wherein it is stated "That the deaths and injuries resulting from the fire aboard the *Forrestal* on 29 July 1967 were not caused by the intent, fault, negligence or inefficiency of any person or persons embarked in the *Forrestal*." Further, the Commander in Chief U.S. Atlantic Fleet specifically does not concur in Opinion 4 of the *Report* which states "That no improper acts of commission or omission by personnel embarked in *Forrestal* directly contributed to the inadvertent firing of the Zuni rocket from F-4 Number 110."[5]

Admiral Holmes also questioned the accuracy of the Investigation Board's finding that the state of *Forrestal's* material readiness and firefighting and damage control training were acceptable at the time of the fire. He noted that the Inspector General of the US Atlantic Fleet conducted a short-notice evaluation of *Forrestal's* damage control readiness on 10 May 1967. The purpose of this visit was to assess the carrier's ability to maintain watertight integrity, fight fires, and repair damage. The Inspector General found *Forrestal's* damage control readiness to be unsatisfactory, and noted that the damage control parties were disorganized and were not knowledgeable. Admiral Holmes's endorsing letter stated that this information was not included in Rear Admiral Massey's *Investigation Report*, although his board was provided with a copy. The admiral's letter further criticized the investigators for not stating whether the unsatisfactory conditions found by the Inspector General were corrected prior to the conflagration in July.[6]

In his endorsing letter, Admiral Holmes stated that although he was concerned with the high turnover rate of enlisted personnel in operational units, he concurred with Admiral Booth that it would be difficult to stabilize manning. He wrote that the low reenlistment rates cited by Admiral Booth were exacerbated by the Navy's low overall manning of enlisted supervisory personnel (enlisted pay grades E5 to E9 were only manned at 82 percent of allowance in August 1967).[7] Other factors that Admiral Holmes assessed as negatively impacting manning stabilization on ships included high operational tempo to support the Navy's heavy commitment in Southeast Asia, and the need to man a larger fleet as the number of ships that were commissioned and reactivated increased.[8]

Admiral Holmes approved all other damage control related recommendations included in Rear Admiral Massey's *Investigation Report*, and forwarded the report to the Navy's Judge Advocate General. The Judge Advocate General reviewed the investigation report and endorsing letters, found that the investigation had been conducted in accordance with naval regulations, and forwarded the entire package to the Chief of Naval Operations. The Judge Advocate General also sent copies of the report and endorsing letters to the commanders of the Naval Air Systems Command, the Naval Ship Systems Command, the Naval Ordnance Systems Command, and the Chief of Naval Personnel for their information. After the Chief of Naval Operations reviewed the report, it was returned to the Judge Advocate General's office.

When the Chief of Naval Operations returned the original copy of Rear Admiral Massey's investigation into *Forrestal's* fire, the Judge Advocate General's office placed it in their long-term storage facility.[9] It appears that the Navy never tracked the status of recommendations made in this report.[10] Fortunately, all but one of the damage control related recommendations first proposed by Rear Admiral Massey were also included in Admiral Russell's report. The sole recommendation excluded by Admiral Russell was the proposal to stabilize manning on Navy ships from the period of Refresher Training through deployment. Perhaps Admiral Russell omitted it since Vice Admiral Booth and Admiral Holmes had already rejected it as infeasible. In any event, Admiral Russell's recommendations were targeted at improving damage control training without the benefit of manning stabilization.

In contrast, the recommendations proposed by Admiral Russell were tracked very closely for several years, as the remainder of this chapter will show.

Implementation of the *Russell Report's* Recommendations

The scope of Admiral Russell's panel was much broader than the investigation into *Forrestal's* fire, as discussed earlier. The Chief of Naval Operations to appointed Admiral Russell:

> Examine actual and potential causes of fires and explosions in aircraft carriers with object of minimizing their occurrence, limiting injuries and damage that result when they occur, and greatly improving the effectiveness of firefighting capability and the control of explosive damage particularly on the flight deck and in the hangar bays.[11]

Admiral Russell submitted his report to Admiral Moorer, the Chief of Naval Operations, on 16 October 1967. Three days later, Admiral Moorer forwarded the report to an extensive array of naval commanders, including the Atlantic and Pacific Fleet Naval Air Force Commanders, all fleet commanders, all aircraft carrier division commanders, all aircraft carrier commanding officers, the Chief of Naval Material, the Chief of Naval Personnel, Naval Ship Systems Command, Naval Ordnance Systems Command, and the Naval Air Systems Command. Admiral Moorer appointed one of the senior officers on his staff, Rear Admiral Edward C. Outlaw, to coordinate implementation of the recommendations submitted by Admiral Russell.[12] Each of the recommendations included in the Russell Report included a proposal for a designated naval command to assume cognizance for further study and implementation if feasible. Admiral Moorer instructed these commands to provide him with their comments on each of these items by 25 November 1967.[13]

Only one of the seventeen significant damage control recommendations included in Admiral Russell's report and discussed in the previous chapter was quickly rejected as infeasible. The discarded recommendation proposed that the Navy allocate a portion of each ship's operating funds solely for the purchase of damage control items. The prioritization of operating funds was traditionally decided by each ship's commanding officer. The commanding officer was in a better position to understand his ship's requirements than higher headquarters staff officers, and was also responsible for everything aboard his ship--the condition of all equipment and the safety of the crew. Additionally, the operating funds were distributed to ships on a quarter-annual basis. It would be exceedingly difficult for outsiders to predict how much damage control equipment would have to be replaced in a given quarter, since wear and tear varied widely according to the ship's operational tempo, how often the gear was used, and how recently it had been replaced. The Navy's leaders decided to leave responsibility for allocation of damage control funding from operating funds with each ship's commanding officer.[14]

Feedback from the offices charged with studying the feasibility of implementing the recommendations put forth in the *Russell Report* indicated that substantial time would be required to perform the required analysis. As a result, in July 1968 the Chief of Naval Operations directed the Chief of Naval Material to provide quarterly reports updating the status of the proposed recommendations. These quarterly status reports were submitted to the Chief of Naval Operations from 1968 until 1972 and detailed progress made in analyzing and implementing the recommendations.

In August 1972, the Chief of Naval operations relaxed the reporting requirement, directing that progress reports be submitted on a semi-annual basis. The Chief of Naval Operations rescinded the reporting requirement entirely in November 1974, since significant progress had been made in implementing the Russell Report recommendations:

> In view of the considerable progress to date implementing Russell Panel/CASS recommendations, it is considered that the periodic status reports have served their intended function and are no longer necessary on a regularly scheduled basis. . . . Ongoing and open-ended recommendations will continue to be monitored and coordinated as normal NAVMAT management actions.[15]

Although the Navy had made enormous progress in implementing Admiral Russell's recommendations by late 1974, interim status updates to the Chief of Naval Operations showed that financial costs proved to be an enormous obstacle to analyzing and implementing the proposed improvements. To ensure that available funding was applied in the most critical areas, the Chief of Naval Operations assigned a relative priority to each recommendation. Three categories of priority were established. The highest category was termed "urgent"; the second, "priority"; and the lowest, "desirable."

Impact of the *Enterprise* Fire on Russell Panel Recommendations

Soon after the Navy began to seriously study the Russell Panel's recommendations, another serious shipboard fire dramatically underscored the need to improve shipboard damage control and firefighting capability. On 14 January 1969, in a tragic parallel to the *Forrestal* fire, a Zuni rocket accidentally ignited on an F-4 Phantom aircraft staged on the aircraft carrier USS *Enterprise's* flight deck. Twenty-seven sailors perished in the resulting blaze, and 344 others were injured (sixty-five seriously). Damage to the ship was estimated to be just below eleven million dollars and the cost of replacing the fifteen destroyed aircraft and associated aviation equipment was estimated to be approximately 45.5 million dollars.[16] The following day, the Pacific Fleet Naval Air Force Commander directed Rear Admiral Frederic A. Bardshar to investigate the fire. Rear Admiral Bardshar's panel also consisted of two Navy Captains, one Commander, and a Lieutenant. Lieutenant Commander Thomas E. Flynn was assigned to provide legal counsel for the investigating board.

Admiral Bardshar completed his report on 11 February 1969. A brief examination of his report is useful for three reasons – first, because the topic of investigation was a similar fire on an aircraft carrier similar to *Forrestal*. Secondly, since the *Enterprise* fire occurred approximately eighteen months after the conflagration on *Forrestal*, sufficient time had elapsed to determine if any suggested improvements had been implemented. Finally, a section of Admiral Bardshar's report commented directly on his opinions of specific *Russell Report* recommendations, based on his investigation of *Enterprise's* fire.

Admiral Bardshar's investigation revealed that although the majority of recommendations proposed to improve shipboard damage control equipment had not yet been implemented, many of the training deficiencies noted by Admirals Massey and Russell had been corrected. In fact, Admiral Bardshar's report vividly illustrates that Enterprise's crew exhibited high levels of damage control awareness and was well trained in damage control and firefighting. In the abstract to his report, Admiral Bardshar stated that although serious firefighting equipment deficiencies existed, "solid damage control organization, training, and execution" minimized casualties and limited the fire's spread and resulting damage.[17] Admiral Bardshar praised the performance of *Enterprise's* firefighters in his report:

> The high state of training which existed aboard *Enterprise* produced the individual leadership at all levels which is necessary to an effective damage control organization. . . . After each major explosion hose teams regrouped and resumed their efforts. When men fell, trained backup men took their place. In any event, the aggressive but controlled efforts of these fire fighting crews prevented the explosions of more 500 pound bombs which almost certainly would have occurred had the fires been allowed to burn unopposed.[18]

This description presented a stark contrast to firefighting efforts on *Forrestal*, where men with little or no formal training took the place of fire fighters who were killed in the initial explosions on that vessel. On *Forrestal*, approximately 50 percent of the ship's crew and none of the air wing sailors had attended firefighting school. When *Enterprise's* fire erupted, 2,997 of the 3,123 sailors in her ship's company (96 percent) had attended firefighting school, and 1,753 of 2,039 air wing personnel (86 percent) had attended firefighting school. *Enterprise* had sent 1,091 officers and men to firefighting school during August and September 1968. The carrier also had developed a damage control training team to instruct and evaluate the performance of its damage control organization during drills. *Enterprise* had also established a competitive program between its repair parties to increase effectiveness, and conducted frequent training drills.[19] Clearly, on *Enterprise* at least, the importance of an effective, highly trained damage control organization was well recognized.

In the portion of his report commenting on the Russell Panel's recommendations, Admiral Bardshar generally concurred with the proposed solutions. He concurred with the first recommendation included in Admiral Russell's report, the need to develop an advanced flight deck fire fighting system for carriers. Admiral Bardshar wrote that although Enterprise's well-trained crew quickly employed all available firefighting equipment in accordance with sound, prescribed doctrine, the firefighting equipment was simply insufficient. As a result, the crew's efforts failed to prevent ordnance cook-off and the significant damage resulting from these explosions. These comments on the *Enterprise* fire were an almost identical echo to those made seventeen months earlier by Rear Admiral Massey. Admiral Bardshar wrote that an advanced flight deck fire system, originally proposed by Rear Admiral Massey, and further endorsed by Admiral Russell, was badly needed. He made this his foremost recommendation, and defended his rationale in the strongest terms:

> A fresh concept of dealing with a massive flight deck fire (whether self or enemy inflicted) involving exploding fuel and ordnance should be developed. The system derived must include massive cooling as well as rapid extinguishment. It must provide flexibility, selectivity, and redundancy. The system must not compete with other systems for power, water, or extinguishing agents. Controls must provide for remote activation and response must be immediate. . . . The requirement for this system is documented by 161 lives, some 200 million dollars, and the loss of 8 CVA months of operating time since 29 July 1967. The system should be a military characteristic for all CVAs and rank in importance with the armament and aircraft launch and recovery systems…Anything less will not be satisfactory.[20]

Admiral Bardshar also agreed that the Navy needed most of the improvements proposed in the *Russell Report*. He opined that a standardized marking and lighting system for escape routes would be desirable, as would the neoprene hoses described by Admiral Russell. At the time of *Enterprise's* fire, the improved neoprene hoses were approved for use on naval vessels. However, the *Enterprise* was not yet fitted with them. Admiral Bardshar also wrote that although improvements to OBAs would be desirable, he felt that improved training (and the resulting increased familiarity sailors had with the equipment's capabilities and limitations) had alleviated many of the perceived shortcomings of OBAs. Admiral Bardshar's panel wrote that the *Enterprise's* crew members were aware of the limitations of using the gas masks as escape breathing devices, and effectively used the gas masks during the blaze. The *Enterprise* fire investigators did agree that improved personnel protective equipment was needed. They noted that two sailors wearing aluminized proximity suits were injured after the hoods were blown off their suits by the concussion from explosions on the flight deck. They also stated that more fire resistant clothing and use of gloves would have reduced the severity and number of burns suffered by *Enterprise's* firefighters. They recommended that the Navy issue and require all personnel working on flight decks to wear hard shell helmets and gloves.[21]

The only *Russell Report* recommendation Admiral Bardshar's investigators disagreed with was the need to increase the allowance of OBA canisters and containers of foam concentrate. *Forrestal* carried 3,300 OBA canisters and 1,220 five-gallon containers of foam concentrate at the time of her fire. *Enterprise's* allowance was virtually identical to this when her fire erupted. Enterprise's crew members expended 900 of their 3,300 OBA canisters and 811 of 1,080 foam concentrate containers while fighting the conflagration.[22] In view of this, Admiral Bardshar wrote that the existing allowance for these items was adequate.

The *Enterprise* investigation indicated that the Navy had made substantial progress in improving personnel training. It also demonstrated that the existing firefighting doctrine was adequate, when used by a highly proficient damage control organization. However, the investigation report also reinforced the assertions contained in the *Forrestal Investigation Report* and the *Russell Report* that existing firefighting and damage control equipment was inadequate. Training had improved human performance, but the Navy's technical experts still had to improve the tools available to shipboard firefighters.

In 1968, the Naval Air Systems Command, operating under authority of the Chief of Naval Material, established the Carrier Aircraft Support Study (CASS). The purpose of the study was to assess aircraft carrier operations, and to recommend improvements to increase effectiveness and safety. CASS was a mammoth study (comprising fourteen volumes; the volume on safety alone contained over 500 pages), and examined nearly every aspect of aircraft carrier operation. The Navy contracted Systems Associates, Incorporated (SAI) to perform the study. SAI subcontracted several major defense-related corporations to provide technical assistance and analysis. Some of the subcontractors who contributed to CASS were FMC Corporation, Grumman Aerospace, Hughes Aircraft, McDonnell Aircraft, and the Western Gear Corporation.[23]

In February 1969 the Chief of Material, acting with the concurrence of the Chief of Naval Operations, directed that follow-up study of recommendations resulting from the *Enterprise* fire be assigned to CASS:

> The recent *Enterprise* incident indicates lessons learned from *Forrestal* contributed to minimizing damage. CASS has been reoriented with OPNAV concurrence to give top priority to *Enterprise*. Coordinated follow-up of *Enterprise* for both short and long term necessary actions are now assigned to CASS. The CASS Steering Committee has been augmented by 2 Flag Officers from OPNAV (OP-03V and OP-50) and the working group is being expanded.[24]

Since several damage control recommendations included in the *Enterprise Investigation Report* were originally included in the *Russell Report*, this action increased the attention accorded to important recommendations that had not yet been implemented. It also provided funding for those recommendations, such as the advanced flight deck firefighting system, that needed significant research and analysis prior to development. The Chief of Naval Material also included the status of recommendations assigned to CASS for further study in the periodic update of *Russell Report* recommendations to the Chief of Naval Operations.[25]

A review of these periodic updates on the status of analysis and implementation of Russell Report recommendations shows that steady progress was made. For instance, by January 1971, fifty separate SHIPALTS (alterations designed to improve Navy ships) based on improvements recommended by Admiral Russell had been approved.[26] Perhaps the most important of these new SHIPALTS was a newly designed Advanced Flight Deck Fire Fighting System for aircraft carriers. However, SHIPALTS had also been developed to improve shipboard "1MC" general announcing systems and increase the number of exits from carrier working and berthing spaces. Unfortunately, the cost of altering the Navy's ships was high, and some SHIPALTS other than those developed from Russell Report recommendations were given higher priority.[27] The May 1971 status update to the Chief of Naval Operations stated that the two aircraft carriers that were being constructed (USS *Nimitz* and USS *Eisenhower*) would have the new damage control improvements built into them, at an estimated additional cost to the Navy of five million dollars per ship.[28] According to that document, approximately $21.5 million were required to complete the fifty SHIPALTS generated by *Russell Report* recommendations on the Navy's existing ships. The Navy had budgeted approximately $13.2 million for this over the next five fiscal years, leaving an unfunded shortfall of approximately $7.3 million.[29] The same report stated that a shortage of research and development funds had slowed implementation of several other important *Russell Report* recommendations. The most significant of these affected recommendations were standardized marking of escape routes from shipboard compartments, development of an emergency escape breathing device, and OBA improvement. The report stated that the Chief of Naval material had requested $4.25 million for research and development of these items in fiscal years 1970 through 1972, but was only granted $2.8 million.[30]

The following year, on 29 October 1972, a machinery space fire in the aircraft carrier USS *Saratoga* killed three sailors and injured twelve others. The deaths were caused by smoke inhalation, and the injuries consisted of burns and smoke inhalation. On 1 November 1972, the Chief of Naval Operations directed his staff to provide him with a status report on the development of *Russell Report* recommendations.[31]

The November 1972 update revealed substantial additional progress on many *Russell Report* recommendations, including the three that had been funded at lower levels than requested the previous year. The Chief of Naval Material had completed evaluation of a standardized marking and lighting system for shipboard escape routes, and was preparing the specifications needed to create a SHIPALT. Research, development, testing, and evaluation (RDT&E) had also been completed on an improved "Variable-fog" nozzle for Navy firefighting hoses. Specifications for the new nozzle were complete, and the Navy was preparing to purchase and equip its ships with them. An emergency escape breathing device had also been developed. This device provided shipboard personnel with eight minutes of clean breathing air to allow them to escape smoke-filled compartments. The Navy had awarded a contract for production of these devices, and was expecting them to be delivered to its ships by late 1973. The report also noted that a permanent flight deck personnel protective equipment program had been established by the Naval Air Systems Command, and that testing of Nomex fire retardant clothing was in progress. Finally, the 1972 status report described an improved OBA that was being evaluated and refined.[32]

The Chief of Naval Material published the final status report on Russell Report recommendations in March 1974. This update showed that, although many research and development efforts were still underway, the Navy had made enormous overall progress in implementing the *Russell Report* recommendations. An advanced flight deck fire fighting system had been installed in nine aircraft carriers, and installation was expected to be completed on the seven remaining carriers by late 1974. A SHIPALT was authorized to standardize shipboard escape route marking, and funding was allocated for ten carriers to receive the alteration in fiscal year 1974. A SHIPALT to improve the "1MC" general announcing system was funded for all Navy ships. SHIPALTS were funded to improve exits from carrier working and berthing spaces. Four carriers were equipped with newly developed emergency escape breathing devices, and funding was allocated for further refinement of these devices. Funding was allocated to replace all of the Navy's OBAs with an improved model over a three-year period. Improved proximity suits were being provided to carriers, although development of improved, fire retardant clothing for sailors was still in progress. Finally, a training film incorporating footage of the *Forrestal* fire had been issued to all Navy fire fighting schools.[33]

This chapter has shown that the vast majority of damage control improvements first proposed by Rear Admiral Massey were eventually implemented, particularly those that called for more effective equipment. Dramatic improvements are difficult to quickly accomplish in a large bureaucratic organization, but several important factors fostered improved damage control throughout the US Navy. Admiral Russell endorsed Admiral Massey's recommendations, and the high degree of interest exhibited by the Chief of Naval Operations helped sustain the necessary resources required to evaluate and implement the recommended improvements. Finally, fires on the carriers *Enterprise* and *Saratoga* underscored the vital, continuing need for the proposed improvements.

The final chapter examines the lasting impact the *Forrestal* fire had on US Navy shipboard damage control, and what implications this fire and its aftermath have for damage control today.

---

[1] Department of the Navy, *Manual of the Judge Advocate General Basic Final Investigative Report Concerning the Fire on Board the USS Forrestal (CVA-59) on July 29, 1967.* (Washington, D.C.: US Navy Office of the Judge Advocate General, 1968), 112, 125.

[2] Vice Admiral Charles T. Booth, *First Endorsement on RADM F. Massey, USN letter of 19 September 1967* (Norfolk, VA: Commander Naval Air Force, US Atlantic Fleet, 26 September 1967), 1.

[3] *Manual of the Judge Advocate General Basic Final Investigative Report Concerning the Fire on Board the USS Forrestal (CVA-59) on July 29, 1967*, 114, 126.

[4] Vice Admiral Booth, 2.

[5] Admiral Ephraim P. Holmes, *Second Endorsement on RADM F. Massey, USN letter of 19 September 1967* (Norfolk, VA: Commander in Chief, US Atlantic Fleet, 1 December 1967), 3.

[6] Ibid., 5-6.

[7] Ibid., 6.

[8] Ibid.

[9] A letter from the Chief of Naval Operations to the Judge Advocate General, dated 21 August 1969, states that Rear Admiral's Massey's *Investigative Report* was returned to the Judge Advocate General on that date. A letter to the author from the Judge Advocate General's office on 23 September 2003 stated that the report is still held in their long-term storage facility in Suitland, Maryland.

[10] The author could not locate any evidence that recommendations proposed in Rear Admiral Massey's *Investigative Report* were tracked by the Navy after the report was given to the Judge Advocate General for safekeeping. Fortunately, many of these recommendations were included in Admiral Russell's report, which was tracked closely for several years.

[11] Naval Safety Center, *Survey of Selected Aircraft Carrier Accidents* (Washington, D.C.: US Naval Safety Center, 1971), 30.

[12] Admiral James S. Russell, *Report of the Panel to Review Safety in Carrier Operations* (Washington, D.C.: Office of the Chief of Naval Operations, 1967), cover letter.

[13] Ibid.

[14] Chief of Naval Material, *Milestone Schedule and Status Report for Implementing the Recommendations of the Russell Panel Report and the Carrier Aircraft Support Study (CASS)* (Washington, D.C.: Chief of Naval Material, 20 March 1974), 3.

[15] Chief of Naval Material, *Periodic Status Reports on Russell Panel/CASS Report* (Washington, D.C.: Chief of Naval Material, 18 November 1974), 1.

[16] Rear Admiral Frederick A. Bardshar, *Record of Proceedings: Formal Board of Investigation Convened by Order of Commander Naval Air Force United States Pacific Fleet to Inquire into the Circumstances Surrounding a Fire Which Occurred on Board USS Enterprise (CVAN 65) on 14 January 1969 Ordered on 15 January 1969* (San Francisco, CA: Rear Admiral Bardshar, 11 February 1969), 21-22.

[17] Ibid., 1.

[18] Ibid., 37.

[19] Ibid., 26.

[20] Ibid., 2, 38.

[21] Ibid., 38-39.

[22] Ibid., 27.

[23] Systems Associates, *Final Report: Carrier Aircraft Support Study (CASS)* (Long Beach, CA: Systems Associates, December 1971), ii-iii.

[24] Chief of Naval Material, *Carrier Aircraft Support Study (CASS) Enterprise Responsibilities, Assignment of* (Washington, D.C.: Chief of Naval Material, 3 February 1969), 1.

²⁵Chief of Naval Material, 18 November 1974, 1.

²⁶Chief of Naval Material, *Budgetary and Cost Summary Russell Panel/CASS Recommendations* (Washington, D.C.: Chief of Naval Material, 17 May 1971), 1-3.

²⁷Ibid.

²⁸Ibid., 4.

²⁹Ibid.

³⁰Ibid., 5.

³¹Office of the Chief of Naval Operations, *RDT&E Efforts Associated with the Russell Panel Report* (Washington, D.C.: Office of the Chief of Naval Operations, November 1972), 1. Research at the Naval Sea Systems Command headquarters at the Washington Navy Yard revealed a memo from the Chief of Naval Operations requesting this special update "in view of the *Saratoga* fire."

³²Ibid., 5-8.

³³Chief of Naval Material (20 March 1974), 4-8.

CHAPTER 6

CONCLUSION

<u>Lasting Impact of the *Forrestal* Fire</u>

This thesis examined what lessons the Navy learned in the area of damage control from the July 1967 fire on USS *Forrestal*, and how the Navy applied these lessons to improve fleetwide damage control capability (doctrine, warship construction features, and damage control equipment). The research has demonstrated that the damage control capability of US Navy ships was significantly improved as a direct result of lessons learned from the July 1967 fire on USS *Forrestal*. Significant changes in the area of damage control resulted from analysis of this disaster, and these changes had lasting positive impact on US Navy damage control capability.

The tremendous loss of life, high number of injured sailors, extensive property damage to the ship and its complement of aircraft, and the loss of several months of operating time for a capital ship captured the attention of the Navy's top leaders. These leaders ordered a thorough investigation into the *Forrestal* fire. Although the resulting 7,500-page report highlighted several serious deficiencies in *Forrestal's* damage control capabilities, the scope of Rear Admiral Massey's investigation was necessarily limited. The Chief of Naval Operation's appointment of retired Admiral James Russell to review safety of aircraft carrier operations throughout the Navy had a much greater impact on improving damage control throughout the fleet. Admiral Russell found that most of the deficiencies found by the *Forrestal* fire investigators also existed aboard the Navy's other aircraft carriers. As a result, Admiral Russell incorporated all but one of Admiral Massey's thirty-one damage control improvement recommendations into his own report.

Lasting Impact on Doctrine

The fire had a relatively minor impact on damage control doctrine, which was fundamentally sound. The Navy's damage control doctrine had evolved with its ships over the years, and incorporated hard-learned lessons from earlier fires and battle damage sustained by Navy vessels.

However, Rear Admiral Massey and Admiral Russell discovered that the damage control proficiency of US Navy aircraft carrier crews was low because of inadequate training. For example, only 50 percent of *Forrestal's* crew members, and none of the embarked air wing personnel (who comprised approximately 40 percent of the sailors aboard *Forrestal*) had completed fire fighting training courses.[1] Admiral Russell wrote that the Navy's existing damage control training requirements were not being met because of insufficient damage control school capacity, high personnel turnover, and the low priority given to damage control readiness by many aircraft carrier commanding officers.[2] Poorly trained sailors were simply not able to competently fight serious fires in accordance with established doctrine.

These training deficiencies were relatively easy to correct in a short period of time. Damage control training facilities were expanded, and senior leaders directed Commanding Officers to ensure that their crews were properly trained. Rear Admiral Massey's investigation report into the *Forrestal* fire was widely distributed throughout the fleet. All of these measures increased damage control awareness throughout the fleet, at least in the short term. The similar fire on USS *Enterprise* nearly eighteen months later provided evidence that many training deficiencies had been corrected. The investigation report into the *Enterprise* fire praised crew members for efficiently fighting the conflagration in accordance with prescribed doctrine to minimize damage.[3] However, this report highlighted the Navy's dire need for the improvements in damage control and personnel protective equipment proposed in Admiral Russell's report.

Lasting Material Impact

Much of the long-term impact of the *Forrestal* fire can be found by examining the improved material items (warship construction features, damage control and personnel protective gear) that were proposed and developed in response to lessons learned from that event. These important developments were built into newly constructed vessels, and many existing ships were altered to incorporate the new technology. Refined versions of this equipment can be found on today's naval warships.

Successful material achievements included development of an advanced flight deck firefighting system, improved personnel protective equipment (including fire retardant uniforms, emergency escape breathing devices, and improved OBAs), improved hoses and nozzles. Navy officials also approved a standardized marking and lighting system for escape routes from interior compartments, and additional exits were constructed for many of these interior compartments.

Like the proposed training improvements, these material improvements were also relatively easy for the Navy to implement. Admiral Russell had access to the Navy's top military and civilian experts while developing his recommendations and substantial evidence indicated that they were necessary. The senior officer in the US Navy, Admiral Moorer, demanded frequent updates on the status of implementing *Russell Report* recommendations. As a result, there was little controversy over and broad support among the Navy's leadership for the vast majority of these proposed material improvements.

These recommendations were also prioritized to meet funding limitations. The January 1969 fire aboard *Enterprise* provided additional evidence of the validity of the proposed improvements. Although funding constraints, research, development, and testing all slowed implementation of these recommendations, the most significant recommendations were all incorporated into US Navy ships within a few years.[4]

Unsuccessful Damage Control Improvement Ideas

The preceding paragraphs have shown that training deficiencies and material deficiencies were rectified relatively easily. There was ample evidence that these deficiencies existed, and clear-cut solutions were readily developed to mitigate them. Most of the proposed solutions were noncontroversial, and enjoyed broad support from senior Navy leaders. However, recommendations that did not have such clear-cut technical solutions and challenged existing policies and organizational culture proved much more difficult to successfully implement.

Three significant recommendations proposed to improve shipboard damage control readiness in the wake of the *Forrestal* fire never materialized. Rear Admiral Massey proposed that the Bureau of Naval Personnel should stabilize manning of trained personnel on ships and air wings by minimizing personnel transfer from these units prior to deployment.[5] However, the two senior admirals who endorsed his report prior to its submission to the Chief of Naval Operations rejected this proposal, primarily because of low manning levels at that time. Admiral Moorer did not insist that his subordinates find a way to stabilize manning. This recommendation was the one significant damage control improvement recommendation first proposed by the *Forrestal* fire investigators that Admiral Russell did not include in his report. Perhaps Admiral Russell sensed or was told that manning stabilization was not feasible during his interviews with the senior officers who rejected the concept after Admiral Massey first proposed it. Admiral Russell's report did include several recommendations designed to ameliorate damage control training proficiency without manning stabilization. These proposals included increasing the emphasis on damage control training for officers and enlisted personnel prior to reporting to their first ships, and increasing the capacity of the fleet damage control training schools. Senior Navy leaders quickly accepted these alternative proposals. Still, Admiral Russell's failure to recommend manning stabilization reduced the visibility of this proposal.

Admirals Massey and Russell both recommended increasing the number of OBAs, OBA canisters, and containers of firefighting foam concentrate carried aboard Navy ships, citing shortages of these items during the *Forrestal's* fire. Vice Admiral Booth objected to immediately implementing this proposal, writing that additional analysis was required before dedicating additional funding and limited shipboard storage areas to these items.[6] In his investigation report on the *Enterprise* fire, Rear Admiral Bardshar flatly rejected the need for additional quantities of these items, writing, "the *Enterprise* allowance for OBAs, canisters, foam, fire extinguishers and hoses was adequate."[7] It appears likely that *Enterprise's* crew used less of these items in a fire very similar to that on *Forrestal* due to their higher training proficiency. In any event, the conflicting data on whether additional quantities of these items were actually required appears to have shifted the focus of Navy leaders to other recommendations with broader support.

The final significant recommendation not implemented by Navy leaders, dedicated funding for replacement of damage control items, was proposed solely by Admiral Russell. The *Russell Report* noted that damage control funding competed with all of the other requirements each ship had, and asserted that many Commanding Officers failed to place a high priority on damage control equipment. Admiral Russell wrote that this frequently resulted in poor material condition of damage control gear.[8]

Although Admiral Russell's logic was sound, this recommendation did not mesh well with Navy culture and tradition. Navy commanding officers were traditionally given complete authority to decide how to allocate limited operational funding for their ship. Many valid reasons existed for this arrangement--commanding officers were held completely responsible for the safety of the ship and its crew. Commanding officers also were presumed to have a much more intimate picture of their ship's condition and requirements, and were thus in a better position to determine allocation of operational funding than outsiders were. It appears likely that senior Navy officials were unwilling to take this decision-making authority away from commanding officers, or allocate additional dedicated funding for damage control items.[9]

<u>Implications for Today's Navy</u>

In July 1967 many people in the Navy thought that a flight deck fire on the magnitude of that on *Forrestal* was unlikely to occur. It was easy for them to believe that technological innovations such as armored flight decks and replacement of highly flammable aviation gasoline with less flammable jet fuel significantly reduced the risk of serious fire. However, the *Forrestal's* fire demonstrated that fire at sea remains a serious and enduring threat to the safety of ships and sailors.

Forrestal's designers built a ship that carried more aircraft, fuel, and ordnance then any earlier aircraft carriers. Unfortunately, the July 1967 fire on *Forrestal* provided strong evidence that these designers failed to ensure that her damage control capability was adequate for these increased hazards. The tragedy illustrated the vital, continuing need to assess damage control capability in new ship designs.

The fires on the *Forrestal* and *Enterprise* also demonstrated the importance of a well-trained and equipped damage control organization. The investigation reports into those incidents provide strong evidence that many sailors died needlessly on *Forrestal* because of poor training. Although it is true that their damage control equipment was inadequate, the fact remains that most sailors aboard *Forrestal* were not trained to effectively use the tools available to them. Conversely, *Enterprise's* well-trained crew was able to effectively fight a similar fire in January 1969, when the events on *Forrestal* were still very fresh in the minds of Navy personnel.

The events following the *Forrestal* fire also provide useful insight into one way to successfully implement change in a large, bureaucratic organization. The tremendous loss of life and high property damage certainly provided a sharp warning that the status quo of damage control on aircraft carriers was inadequate. Senior Navy leaders acted decisively to improve this situation. The Navy's senior officer appointed a retired four-star admiral to head a panel tasked with examining the safety of aircraft carrier operations. This officer, Admiral James Russell, was granted unfettered access to the Navy's top ship construction and damage control experts and the most experienced naval officers while developing proposals to improve damage control readiness. The Chief of Naval

Operation's strong personal commitment to the project sustained momentum throughout the several years required to implement the proposed solutions.

---

[1] Admiral James S. Russell, *Report of the Panel to Review Safety in Carrier Operations* (Washington, D.C.: Office of the Chief of Naval Operations, 1967), A-73--A-74.

[2] Ibid., A-74-A-76.

[3] Rear Admiral Frederick A. Bardshar, *Record of Proceedings: Formal Board of Investigation Convened by Order of Commander Naval Air Force United States Pacific Fleet to Inquire into the Circumstances Surrounding a Fire Which Occurred on Board USS Enterprise (CVAN 65) on 14 January 1969 Ordered on 15 January 1969* (San Francisco, CA: Rear Admiral Bardshar, 11 February 1969), 1-2, 26.

[4] Chief of Naval Material, *Milestone Schedule and Status Report for Implementing the Recommendations of the Russell Panel Report and the Carrier Aircraft Support Study (CASS)* (Washington, D.C.: Chief of Naval Material, 20 March 1974), 1-8.

[5] Department of the Navy, *Manual of the Judge Advocate General Basic Final Investigative Report Concerning the Fire on Board the USS Forrestal (CVA-59) on July 29, 1967* (Washington, D.C.: US Navy Office of the Judge Advocate General, 1968), 125.

[6] Vice Admiral Charles T. Booth, *First Endorsement on RADM F. Massey, USN letter of 19 September 1967* (Norfolk, VA: Commander Naval Air Force, US Atlantic Fleet, 26 September 1967), 2.

[7] Rear Admiral Bardshar, 39.

[8] Admiral Russell, A-33.

[9] Chief of Naval Material document of 20 March 1974, page 3, simply states that this proposal was rejected as "not feasible." A search of the Navy Operational Archives and records at the Naval Sea Systems Command at the Washington Navy Yard failed to provide any further details on why this proposal was considered infeasible.

APPENDIX A

TYPICAL NAVY FIREMAIN "LOOP" DIAGRAM

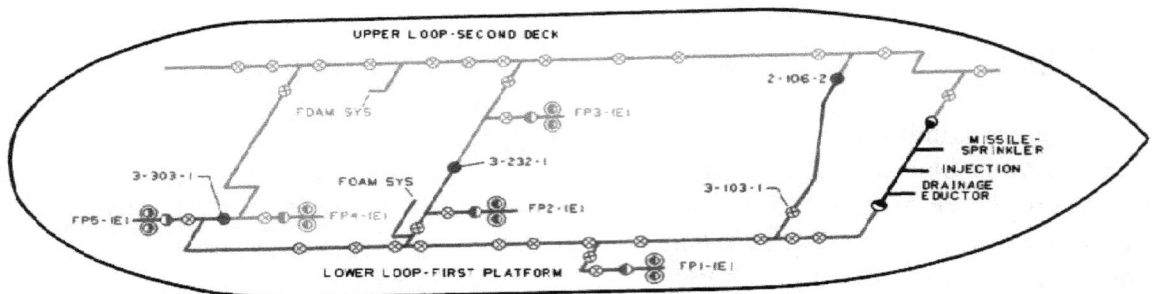

ALL PUMPS AVAILABLE FOR OPERATION OF THE MK13 GUIDED MISSILE LAUNCHING SYSTEM, SPRINKLER SYSTEM, BOOSTER SUPPRESSION & DRAINAGE EDUCTORS

LOWER LOOP - FP1-(E), FP2-(E), & FP5-(E)

UPPER LOOP - FP3-(E), & FP4-(E)

THE FIREMAIN LOOP IS SEGREGATED INTO TWO SECTIONS. IN CONDITION "ZEBRA" AS INDICATED ABOVE. TO ESTABLISH CONDITION "ZEBRA" FROM CONDITION "X-RAY" OR "YOKE", THE FOLLOWING THREE "ZEBRA" VALVES ARE CLOSED.

VALVE NO.	LOCATION	REMOTE CONTROL
2-106-2	2-100-4-L	2-292-01-C
3-232-1	5-212-0-E	2-292-01-C
3-303-1	5-292-0-E	2-292-01-C
3-103-1	3-100-1-L	

Cutaway View of Loop

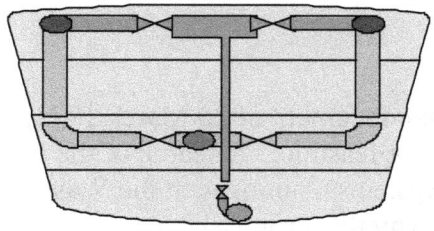

Source: These diagrams originally appeared in a training presentation prepared by the US Navy's Surface Warfare Officer School at the Naval Education and Training Center, Newport, Rhode Island.

APPENDIX B

CHRONOLOGY OF EVENTS

This timeline was developed using the official Navy records in the Bibliography as references.

October 1955 – USS *Forrestal (CVA-59)* was commissioned.

April 1966 – January 1967 – *Forrestal* docked in Norfolk Naval Shipyard for extended maintenance and overhaul.

May 1966 – Captain John K. Beling assumed command of *Forrestal*.

January – May 1967 – *Forrestal* conducted predeployment training.

6 June 1967 – *Forrestal* departed Norfolk for deployment to Western Pacific.

24 July 1967 – *Forrestal* arrived on Yankee Station, Gulf of Tonkin, off coast of North Vietnam.

25 – 28 July 1967 – *Forrestal* launched air strikes against targets in North Vietnam.

29 July 1967 – A Zuni rocket accidentally fired from an F-4 Phantom jet staged on *Forrestal's* flight deck struck a nearby A-4 Skyhawk aircraft and started a large conflagration aboard the ship at 1051 local time. 134 personnel died; 161 others were injured.

30 July 1967 – The fires aboard *Forrestal* were extinguished by 0400 local time. Rear Admiral Lanham, commander of Carrier Division Two (Captain Beling's Immediate Superior in Command), who was embarked in *Forrestal* during the fire, ordered his staff to conduct a preliminary investigation into the fire. Vice Admiral Charles T. Booth, commander of the US Atlantic Fleet Naval Air Force, appoints Rear Admiral Forsyth Massey as senior member of the board of investigation into the *Forrestal* fire.

31 July 1967 – *Forrestal* arrived at Naval Air Station Cubi Point, Republic of the Philippines, for temporary repairs.

3 August 1967 – Rear Admiral Massey and his team of investigators arrived at Naval Air Station Cubi Point and began work. Rear Admiral Lanham's preliminary investigation ended; all information gathered was turned over to Admiral Massey's team.

21 August 1967 – Admiral Thomas Moorer, Chief of Naval Operations, appointed retired Admiral James S. Russell as director of a panel charged with reviewing safety in aircraft carrier operations.

19 September 1967 – Rear Admiral Massey completed his investigation and forwarded his report to Vice Admiral Booth.

26 September 1967 – Vice Admiral Booth completed his review of Rear Admiral Massey's report and forwarded it to Admiral Ephraim P. Holmes, Commander in Chief of the US Atlantic Fleet.

16 October 1967 – Admiral Russell completed his *Final Report of Panel to Review Safety in Carrier Operations* and submitted it to the Chief of Naval Operations.

19 October 1967 – Chief of Naval Operations forwarded Admiral Russell's report to an extensive list of naval commands, assigned Rear Admiral Edward C. Outlaw to coordinate analysis and implementation of proposed recommendations, and directed subordinate commands to provide comments on the proposed recommendations by 25 November 1967.

1 December 1967 – Admiral Holmes completed his review of Rear Admiral Massey's report.

23 July 1968 – Chief of Naval Operations directed the Chief of Naval Material to provide quarterly updates on the status of *Russell Report* recommendations.

November 1968 – The Naval Air Systems Command established a Carrier Aircraft Support Study (CASS) group

14 January 1969 – Flight deck fire erupted on the aircraft carrier *USS Enterprise* after a Zuni rocket exploded while attached to an F-4 Phantom jet staged on deck. Twenty-seven personnel perished; 344 others were injured (65 seriously).

15 January 1969 – Commander of US Pacific Fleet Naval Air Force appointed Rear Admiral Frederic A. Bardshar to investigate *Enterprise* fire.

February 1969 – Rear Admiral Bardshar completed his report. His report validated the necessity of nearly all of the damage control improvements proposed by Rear Admiral Massey and Admiral Russell. Chief of Naval Operations directed the CASS group to focus on following up lessons learned from Enterprise fire. The status of these recommendations were included in future quarterly progress reports to the Chief of Naval Operations outlining progress on *Russell Report* recommendations.

21 August 1969 – The Chief of Naval Operations returned Rear Admiral Massey's report, with endorsing letters from Vice Admiral Booth and Admiral Holmes, to the Navy's Judge Advocate General for storage.

December 1971 – Carrier Aircraft Support Study (CASS) completed.

28 August 1972 – Chief of Naval Operations directed the Chief of Naval Material to provide semiannual updates (instead of quarterly reports) on the status of *Russell Report* recommendations.

29 October 1972 – Machinery space fire in the aircraft carrier USS Saratoga killed three sailors and injured twelve others. Three days later, the Chief of Naval Operations requested a special report updating him on the status of *Russell Report* recommendations.

18 November 1974 – Periodic status reports on *Russell Report* recommendations discontinued.

APPENDIX C

SUMMARY OF SELECTED *RUSSELL REPORT* RECOMMENDATIONS

Admiral Russell's report was used as the source document for this appendix, which summarizes several of the most significant damage control improvement proposals discussed in the body of the thesis.

Recommendation 1-1: Advanced Flight Deck Fire Fighting System. Proposed features included "remote control, massive and quick response, cooling for ordnance, sufficient redundancy to compensate for derangement of portions of the system...a means for quick drainage or dispersal of large quantities of spilled fuel from the flight deck."

Recommendation 1-6: Marking of Escape Routes. Proposed establishing a "standardized system of marking and lighting emergency escape routes in aircraft carriers" to aid personnel attempting to escape smoke-filled interior compartments.

Recommendation 1-9: Improved Interior Communications. Recommended that all aircraft carriers be surveyed to determine adequacy of the shipboard general announcing system, the "1MC." It also recommended prompt correction of any deficiencies that were discovered.

Recommendation 1-10: Improved Fire Hose. Recommended that the Navy require neoprene wrapped hoses on flight and hangar decks to reduce kinking that commonly occurred with standard cotton-jacketed fire hoses used aboard Navy ships. It also proposed development of quick-disconnect couplings for these hoses.

Recommendation 1-11: Review of Ship Alterations Affecting Safety. Proposed that the Navy review all pending ship alterations and ship repair requests, and that items affecting damage control and firefighting be given high priority during maintenance periods.

Recommendation 1-22: Damage Control Equipage Allowance. Recommended further analysis of fires on the carriers *Oriskany* and *Forrestal* to determine an appropriate allowance for OBAs and their canisters, firefighting foam, fire extinguishers, hoses, and other damage control equipment.

Recommendation 1-23: Funding for Damage Control Equipment. Proposed that the Navy provide ships with dedicated funding for damage control items, to "avoid having safety equipment compete with all other ship upkeep items for the limited funds available."

Recommendation 1-26: Escape Criteria. Proposed changing ship construction criteria to require two exits from berthing compartments and working areas designed for ten or more men. Recommended modifying existing ships to meet these criteria, where feasible.

Recommendation 2-1: Current Mk-V Gas Mask Capabilities. Recommended distributing information to the fleet on the capabilities and limitations of using gas masks as an escape breathing device. The gas mask could be used to filter out airborne particles (protecting the wearer against some contaminants found in smoke), but provided the user with no protection against high levels of carbon monoxide or low oxygen levels.

Recommendation 2-3: Emergency Breathing Apparatus. Proposed development of masks with a small portable oxygen supply to eliminate one of the most serious limitations of using the gas mask as an escape breathing device.

Recommendation 2-6: Flight Deck Personnel Equipment. Proposed development of more effective personnel protective gear, such as fire retardant clothing.

Recommendation 2-7: OBA Improvement. Advocated further development of OBAs to make them smaller, more robust, and simpler to use.

Recommendation 2-8: Improved Proximity Suit. Recommended development of a more effective proximity suit. Also proposed including specialized boots as an integral part of the new suit.

Recommendation 5-1: Air Wing Damage Control / Fire Fighting Training. Recommended that all air wing personnel receive basic damage control and fire fighting training prior to embarking on an aircraft carrier.

Recommendation 5-2: Fleet Damage Control Training Facilities. Recommended expanding these facilities to meet fleet training requirements.

Recommendation 5-5: En Route Damage Control Training for Enlisted Personnel. Proposed mitigating the effect of high personnel turnover by providing training for junior enlisted personnel before they reported to their first ship.

Recommendation 5-7: Increased Emphasis on Damage Control. Recommended stressing the importance of damage control at the Navy's training commands, including Officer Commissioning School, Naval Reserve Officer Training Corps (NROTC) units, and the Naval Academy.

Recommendation 5-8: Training Aids. Advocated development of more effective and realistic damage control training aids, including a film containing actual footage of the *Forrestal* fire.

GLOSSARY

1 MC. Shipwide general announcing system.

Class Alpha Fire. Involved combustible materials such as bedding, books, and clothing.

Class Bravo Fire. Involved flammable liquids such as oils and paint.

Class Charlie Fire. Occurred in electrical equipment.

Class Delta Fire. Occurred when metals such as magnesium ignited.

Compartment Check-Off List. Posted list of all watertight fittings in a shipboard compartment, or interior subdivision.

Fire Bill. Published list posted on US Navy ships to assign specific duties to crew members in the event of a fire.

Firemain Loop. A continuous line of piping containing firefighting water aboard Navy ships. A diagram of a typical loop is included in Appendix A.

Manual of the Judge Advocate General Investigation. Conducted to determine the cause of an accident, and to identify who should be held responsible for resulting damage.

Material Condition Circle X-Ray. A modification of Material Condition X-Ray. Permitted crew members to open certain pre-designated watertight fittings.

Material Condition Circle Yoke. A modification of Material Condition Yoke. Permitted crew members to open certain pre-designated watertight fittings.

Material Condition Circle Zebra. A modification of Material Condition Zebra. Permitted crew members to open certain pre-designated watertight fittings.

Material Condition X-Ray. The lowest degree of watertight integrity on a US Navy ship. Substantially eases crew access to interior compartments, but was rarely set.

Material Condition Yoke. The intermediate degree of watertight integrity on a US Navy ship. Provided a good balance between convenience for crew and ship safety, and was typically set inport or while ships operated in friendly waters.

Material Condition Zebra. The highest degree of watertight integrity on a US Navy ship. Substantially disrupts crew comfort, and is typically set for training, during emergencies, and prior to expected attack.

Operating Target. Funds allocated to individual ships to purchase items such as paint, damage control equipment, paper, and soap.

Oxygen Breathing Apparatus. Portable oxygen-generating protective gear worn by shipboard firefighters to protect them from toxic gases.

Pilot Landing Aid Television. Camera system that recorded events on aircraft carrier flight decks.

PKP extinguishers. Portable dry chemical fire extinguishers used aboard Navy ships.

Ship Alteration. Approved modification of a vessel to correct an identified deficiency.

William fittings. Shipboard fittings marked with a black letter "W". These fittings were vital to ship operation, and were normally kept open regardless of which material condition was set.

BIBLIOGRAPHY

Bardshar, Frederick A. *Record of Proceedings: Formal Board of Investigation Convened by Order of Commander Naval Air Force United States Pacific Fleet to Inquire into the Circumstances Surrounding a Fire Which Occurred on Board USS Enterprise (CVAN 65) on 14 January 1969 Ordered on 15 January 1969*. San Francisco, CA: Rear Admiral Bardshar, 11 February 1969.

Bonner, Kit, and Carolyn Bonner. *Great Naval Disasters: U.S. Naval Accidents in the 20th Century*. Osceola, WI: MBI Publishing, 1998.

Bureau of Naval Personnel. *Principles of Naval Engineering*. Washington, D.C.: United States Navy, 1970.

Burlage, John D. "The *Forrestal* Fire." *Naval Aviation News*, October 1967, 7-14.

Clancy, Tom. *Carrier: A Guided Tour of an Aircraft Carrier*. New York: Berkley Books, 1999.

Chief of Naval Material. *Carrier Aircraft Support Study (CASS) Enterprise Responsibilities, Assignment of*. Washington, D.C.: Chief of Naval Material, 3 February 1969.

Chief of Naval Material. *Budgetary and Cost Summary Russell Panel/CASS Recommendations*. Washington, D.C.: Chief of Naval Material, 17 May 1971.

Chief of Naval Material. *Milestone Schedule and Status Report for Implementing the Recommendations of the Russell Panel Report and the Carrier Aircraft Support Study (CASS)*. Washington, D.C.: Chief of Naval Material, 20 March 1974.

Chief of Naval Material. *Periodic Status Reports on Russell Panel/CASS Report*. Washington, D.C.: Chief of Naval Material, 18 November 1974.

Commander, Naval Surface Force. *COMNAVSURFORINST 3502.1A: Surface Force Training Manual*. Washington, D.C.: United States Navy, 2003.

Commander, Naval Surface Force. *COMNAVSURFLANTINST 3541.1C: Standard Repair Party Manual for Naval Surface Force*. Norfolk, VA: United States Navy, 1991.

Department of the Navy. *Fire Equipment Tests Aboard the CVA-62 Related to Improved Aircraft Carrier Safety*. Washington, D.C.: United States Navy, 1968.

Department of the Navy. *Human Factors Engineering Deficiencies Aboard CVAs*. Washington, D.C.: United States Navy, 1972.

Department of the Navy. *Investigation of Lighting and Directional Signs for Emergency Egress from Ships*. Washington, D.C.: United States Navy, 1968.

Department of the Navy. *Manual of the Judge Advocate General: Basic Final Investigative Report Concerning the Fire on Board the USS Forrestal (CVA 59)*. Washington, D.C.: United States Navy, 1967.

Department of the Navy, Bureau of Weapons. *Summary of Results of Pyrotechnics Portion of Fast Cook-Off Program*. Crane, IN.: United States Navy, 1969.

Freeman, Gregory A. *Sailors to the End*. New York: Avon Books, 2002.

Friedman, Norman. *U.S. Destroyers: An Illustrated Design History*. Annapolis, MD: United States Naval Institute, 1982.

Gillmer, Thomas C. *Modern Ship Design*. 2nd ed. Annapolis, MD: United States Naval Institute, 1986.

Manning, G. C., and T. L. Schumacher. *Principles of Warship Construction and Damage Control*. Annapolis, MD: United States Naval Institute, 1935.

Maritime Administration. *Marine Fire Prevention, Firefighting and Fire Safety*. Washington, D.C.: United States Maritime Administration, 1987.

McCain, John, and Mark Salter. *Faith of My Fathers: A Family Memoir*. New York: Random House, 1999.

McQuaid, Richard W. *A Survey of Aircraft Carrier Accidents from 1951-1967*. Washington, D.C.: Naval Sea Systems Command, 1968.

Naval Safety Center. *Survey of Selected Aircraft Carrier Accidents*. Washington, D.C.: Naval Safety Center, 1971.

Naval Sea Systems Command. *NSTM Chapter 079v3r2: Damage Control: Practical Damage Control*. Washington, D.C.: Naval Sea Systems Command, 2000.

Naval Sea Systems Command. *NSTM Chapter 555v1r9: Surface Ship Firefighting*. Washington, D.C.: Naval Sea Systems Command, 2002.

Naval Sea Systems Command. *NSTM Chapter 9930: Surface Ship Firefighting*. Washington, D.C.: Naval Sea Systems Command, 1986.

Naval Ship Systems Command. *NSTM Chapter 9930: Fire Fighting - Ship*. Washington, D.C.: Naval Ship Systems Command, 1967.

Naval Weapons Laboratory. *Bomb Survivability in Fire Program*. Dahlgren, VA: Naval Weapons Laboratory, 1972.

Navy Department. *Handbook of Damage Control: NAVPERS PUB 1619*. Washington, D.C.: Bureau of Ships, 1945.

Navy Department. *War Damage Report No. 51, Destroyer Report: Gunfire, Bomb, and Kamikaze Damage, Including Losses in Action, 17 October, 1941 to 15 August, 1945*. Washington, D.C.: Bureau of Ships, 1947.

Navy Department. *War Damage Report No. 56, USS Franklin (CV 13): Suicide Plane Crash Damage, Formosa - 13 October, 1944; Bomb Damage, Luzon – 15 October, 1944; Suicide Plane Crash Damage, Samar – 30 October, 1944; Bomb Damage, Honshu – 19 March, 1945*. Washington, D.C.: Bureau of Ships, 1946.

Navy Department. *NSTM Chapter 93: Fire Fighting - Ship*. Washington, D.C.: Bureau of Ships, 1951.

Noel, John V., and Edward L. Beach. *Naval Terms Dictionary*. Annapolis, MD: United States Naval Institute, 1971.

Office of the Chief of Naval Operations. *RDT&E Efforts Associated with the Russell Panel Report*. Washington, D.C.: Office of the Chief of Naval Operations, November 1972.

Rossell, Henry E., and Lawrence B. Chapman. *Principles of Naval Architecture, Volumes One and Two*. New York: The Society of Naval Architects and Marine Engineers, 1939.

Russell, James S. *Report of the Panel to Review Safety in Carrier Operations*. Washington, D.C.: United States Navy, 1967.

Sanders, Michael S. *The Yard: Building a Destroyer at the Bath Iron Works*. New York: HarperCollins, 1999.

Systems Associates. *Final Report: Carrier Aircraft Support Study (CASS)*. Long Beach, CA: Systems Associates, 1971.

Turabian, Kate L. *A Manual for Writers*. 6th ed. Chicago: University of Chicago Press, 1996.

U.S. Army. Command and General Staff College. ST 20-10, *Master of Military Art and Science (MMAS) Research and Thesis.* Ft. Leavenworth, KS: USA CGSC, July 2003.

Williams, Jerome, John J. Higginson, and John D. Rohrbough. *Sea and Air: The Marine Environment.* 2nd ed. Annapolis, MD: United States Naval Institute, 1975.

INITIAL DISTRIBUTION LIST

Combined Arms Research Library
U.S. Army Command and General Staff College
250 Gibbon Ave.
Fort Leavenworth, KS 66027-2314

Defense Technical Information Center/OCA
825 John J. Kingman Rd., Suite 944
Fort Belvoir, VA 22060-6218

LTC Marian E. Vlasak
CSI
USACGSC
1 Reynolds Ave.
Fort Leavenworth, KS 66027-1352

Mr. David Christie
DJMO
USACGSC
1 Reynolds Ave.
Fort Leavenworth, KS 66027-1352

Dr. Jerold E. Brown
CSI
USACGSC
1 Reynolds Ave.
Fort Leavenworth, KS 66027-1352

CERTIFICATION FOR MMAS DISTRIBUTION STATEMENT

1. Certification Date: 18 June 2004

2. Thesis Author: LCDR Henry P. Stewart, USN

3. Thesis Title: The Impact of the USS *Forrestal's* 1967 Fire on United States Navy Shipboard Damage Control

4. Thesis Committee Members: _____
   Signatures: _____
   _____

5. Distribution Statement: See distribution statements A-X on reverse, then circle appropriate distribution statement letter code below:

(A) B   C   D   E   F   X   SEE EXPLANATION OF CODES ON REVERSE

If your thesis does not fit into any of the above categories or is classified, you must coordinate with the classified section at CARL.

6. Justification: Justification is required for any distribution other than described in Distribution Statement A. All or part of a thesis may justify distribution limitation. See limitation justification statements 1-10 on reverse, then list, below, the statement(s) that applies (apply) to your thesis and corresponding chapters/sections and pages. Follow sample format shown below:

EXAMPLE

Limitation Justification Statement	/	Chapter/Section	/	Page(s)
Direct Military Support (10)	/	Chapter 3	/	12
Critical Technology (3)	/	Section 4	/	31
Administrative Operational Use (7)	/	Chapter 2	/	13-32

Fill in limitation justification for your thesis below:

Limitation Justification Statement	/	Chapter/Section	/	Page(s)
_____	/	_____	/	_____
_____	/	_____	/	_____
_____	/	_____	/	_____
_____	/	_____	/	_____

7. MMAS Thesis Author's Signature: _____

STATEMENT A: Approved for public release; distribution is unlimited. (Documents with this statement may be made available or sold to the general public and foreign nationals).

STATEMENT B: Distribution authorized to U.S. Government agencies only (insert reason and date ON REVERSE OF THIS FORM). Currently used reasons for imposing this statement include the following:

    1. Foreign Government Information. Protection of foreign information.

    2. Proprietary Information. Protection of proprietary information not owned by the U.S. Government.

    3. Critical Technology. Protection and control of critical technology including technical data with potential military application.

    4. Test and Evaluation. Protection of test and evaluation of commercial production or military hardware.

    5. Contractor Performance Evaluation. Protection of information involving contractor performance evaluation.

    6. Premature Dissemination. Protection of information involving systems or hardware from premature dissemination.

    7. Administrative/Operational Use. Protection of information restricted to official use or for administrative or operational purposes.

    8. Software Documentation. Protection of software documentation - release only in accordance with the provisions of DoD Instruction 7930.2.

    9. Specific Authority. Protection of information required by a specific authority.

    10. Direct Military Support. To protect export-controlled technical data of such military significance that release for purposes other than direct support of DoD-approved activities may jeopardize a U.S. military advantage.

STATEMENT C: Distribution authorized to U.S. Government agencies and their contractors: (REASON AND DATE). Currently most used reasons are 1, 3, 7, 8, and 9 above.

STATEMENT D: Distribution authorized to DoD and U.S. DoD contractors only; (REASON AND DATE). Currently most reasons are 1, 3, 7, 8, and 9 above.

STATEMENT E: Distribution authorized to DoD only; (REASON AND DATE) Currently most used reasons are 1, 2, 3, 4, 5, 6, 7, 8, 9, and 10.

STATEMENT F: Further dissemination only as directed by (controlling DoD office and date), or higher DoD authority. Used when the DoD originator determines that information is subject to special dissemination limitation specified by paragraph 4-505, DoD 5200.1-R.

STATEMENT X: Distribution authorized to U.S. Government agencies and private individuals of enterprises eligible to obtain export-controlled technical data in accordance with DoD Directive 5230.25; (date). Controlling DoD office is (insert).

www.ingramcontent.com/pod-product-compliance
Lightning Source LLC
Chambersburg PA
CBHW081840170426
43199CB00017B/2797